DAS TIERREICH

Eine Zusammenstellung und Kennzeichnung der rezenten Tierformen

Herausgeber

ROBERT MERTENS (Wirbeltiere)

WILLI HENNIG (Wirbellose)

Schriftleitung
HEINZ WERMUTH

Unter Mitarbeit zahlreicher Spezialisten

„Πάντα ῥεῖ"

„Sine systemate chaos"

Lieferung 98

Lepidoptera Noctuiformes

Agaristidae II

(Ethiopian and Madagascan Species)

Bearbeitet von

Prof. Dr. S. G. Kiriakoff, Gent

Mit 52 Figuren im Text

Walter de Gruyter · Berlin · New York
1977

Das Tierreich	Berlin	Lieferung 98	Seite I—VIII, 1—165	1977

ISBN 3 11 006858 3

Printed in Germany.
Satz: IBM-Composer Walter de Gruyter & Co., Berlin.
Druck: Karl Gerike, Berlin.
Bindearbeiten: Dieter Mikolai, Berlin.

Lepidoptera Noctuiformes

Agaristidae II

(Ethiopian and Madagascan Genera)

Bearbeitet von

Prof. Dr. S. G. Kiriakoff

Systematic Index

Fam. **Agaristidae**

II. Ethiopian Genera

Andrhippurus 4
A. caudequina 5

Apaegocera 6
A. argyrogramma 7
A. aurantipennis 7

Weymeria 8
W. athene 9

Aletopus 10
A. imperialis 11
A. ruspina 11

Heraclia 12
H. geryon 14
H. pardalina 15
H. atriventralis 15
H. jugans 16
H. butleri 16
H. superba 17
H. s. superba 18
H. s. ugandana 18
H. s. basalifasciata 18
H. s. incongruens 18
H. perdix 19
H. africana 19
H. flavisignata 20
H. pallida 20
H. xanthopyga 21
H. terminatis 21
H. atrifusa 22
H. dificiens 22
H. medeba 23
H. contigua 23
H. aisha 24
H. thruppi 24
H. durbanica 25
H. hornimani 25
H. h. hornimani 25
H. h. melanosoma 26
H. h. meridionalis 26
H. hypercompoides 26
H. elongata 27
H. zenkeri 27
H. z. zenkeri 27
H. z. zeodita 28
H. z. niepelti 28
H. karschi 28
H. gruenbergi 29
H. g. gruenbergi 30
H. g. medjensis 30
H. flavipennis 30
H. longipennis 31
H. l. longipennis 31
H. l. bodaensis 31
H. aemulatrix 31
H. a. aemulatrix 32
H. a. campala 32
H. a. suda 32
H. monslunensis 33
H. nigridorsa 33
H. kivuensis 33
H. buchholzi 34
H. annulata 35
H. aurea 35
H. poggei 35
H. catori 36
H. limbomaculata 36
H. mozambica 37
H. vittei 37

Pentelia 38
P. pentelia 39

Rothia 40
R. panganica 41

Chaetostephana 41
Ch. rendalli 43
Ch. r. rendalli 43
Ch. r. angolana 43
Ch. inclusa 43

Hespagarista 44
H. echione 45
H. caudata 46

Syfanoidea 46
S. schencki 47

Paraegocera 48
P. confluens 48

Metagarista 49
M. triphaenoides 50
M. maenas 51
M. m. maenas 51
M. m. disjuncta 52

Polacanthopoda 52
P. tigrina 53
P. t. tigrina 54
P. t. fabricata 54
P. humphreyi 54
P. anthina 55
P. naveli 56

Aethodes 56
A. angustipennis 57

Aegoceropsis 58
A. affinis 58
A. fervida 59
A. f. fervida 60
A. f. fuscicosta 60
A. f. media 60
A. obliqua 61
A. o. obliqua 61
A. o. isogenis 61
A. o. remutata 61
A. geometrica 62
A. rectilinea 62
A. brevivitta 62
A. b. brevivitta 63
A. b. rectilineoides 63
A. ferrugo 63

Mitrophrys 64
M. menete 65
M. m. menete 66
M. m. rubida 66
M. magna 66
M. m. magna 67
M. m. mesia 67
M. m. obliquisigna 67
M. m. abyssinica 67
M. m. kenyamagaribae 67
M. latreillii 68
M. gynandra 68
M. barnsi 69
M. ansorgei 69

Agoma 70
A. trimeni 71
A. t. trimeni 72
A. t. halans 72

Omphaloceps 72
O. triangularis 74
O. daria 74

Epischausia 75
E. dispar 76

Lophonotidia 77
L. nocturna 77
L. melanoleuca 78

Misa 79
M. memnonia 80
M. cosmetica 81
M. costistrigata 81
M. schultzei 81

Schausia 82
S. leona 82
S. l. leona 83
S. l. congoana 83
S. gladiatoria 84
S. coryndoni 84
S. greenawayi 85

Hoplarista 85
H. haemaplaga 86

Paida 86
P. pulchra 87

Pseudopais 88
P. nigrobasalis 89

Brephos 90
B. decora 91
B. festiva 92
B. f. festiva 92
B. f. nigripectus 93
B. nyassana 93
B. ansorgei 93

Crameria 94
C. amabilis 95

Pseudospiris 95
P. paidiformis 97

Choeropais 97
Ch. jucunda 98

Massaga 99
M. maritona 101
M. angustifascia 101
M. monteirona 101

Massagidia 102
M. hersperica 103
M. tenuifascia 104
M. xenia 104

Paratuerta 104
P. marshalli 106
P. featheri 106

Hemituerta 106
H. nana 108
H. mahdi 108

Hypotuerta 109
H. transiens 110

Tuerta 110
T. chrysochlora 111
T. cyanopasta 112

Pseudotuerta 113
P. argyrochlora 113

Tuertella 114
T. rema 115

Charitosemia 116
Ch. albigutta 117
Ch. geraldi 117
Ch. g. geraldi 118
Ch. g. angustata 118

Ovios 118
O. capensis 119
O. c. capensis 120
O. c. septentrionalis 120
O. nealces 120

Leucovis 121
I. alba 122
L. lepta 122

Acantuerta 122
A. thomensis 124
A. ladina 123

III. Madagascan Genera

Rothia 126
R. micropales 127
R. m. micropales 128
R. m. caecata 128
R. powelli 128
R. dayremi 129
R. holli 129
R. cruenta 130
R. rhaeo 130
R. arosa 130
R. hampsoni 131
R. westwoodi 131
R. viossati 132
R. simyra 132
R. watersii 133
R. hypopyrrha 133
R. zea 134
R. distigma 134
R. agrius 134
R. pedasus 135
R. pales 135
R. metagrius 136
R. turlini 136
R. lasti 137
R. nigrescens 137

Pararothia 138
P. gracilis 139
P. camilla 139
P. vieui 139

Pristoceraea 140
P. eriopis 141

Schausiana 141
S. obrysos 142

Nesaegocera 143
N. comorana 144

Arctiopais 145
A. ambusta 146

Arrothia 146
A. bicolor 148
A. gueneeiana 148

Antigodasa 148
A. rufidorsalis 149

Ancarista 150
A. laminifera 151
A. l. laminifera 152
A. l. comorana 152

Musurgina 152
M. laeta 153

Pemphigostola 153
P. synemonistis 155

Introduction

There is little doubt that the origin of the Ethiopian and Madagascan Agaristidae is to be sought in the Oriental faunistic region. Some genera as e.g. *Heraclia*, *Aegoceropsis* and *Mitrophrys*, have undergone but very little morphological change, and they hardly differ from their eastern relatives, viz. *Episteme* and *Aegocera.* On the other hand, there are genera that have diverged morphologically quite considerably, especially some Madagascan genera. On the African continent, there is a very aberrant genus, *Andrhippuris*, lacking the harpe, and placed here among the Agaristidae with some doubt. The remaining genera can be divided into two groups, according to the structure of the antennae – those genera with the antennae more or less dilated towards the extremity being traditionally considered as the "true" Agaristidae. Those with the antennae differently shaped have received different treatment according to the various authors. The most comprehensive treatment is that adopted in the Lepidopterorum Catalogus, part 5, by E. STRAND. HAMPSON, in the Catalogue of the Lepidoptera Phalaenae in the British Museum, parts 3 and 9, was rather restrictive in his judgment, while JORDAN, perhaps the foremost authority on the Agaristidae, occupied an intermediate position. There is little doubt that filiform antennae are plesiomorphic, but ciliate, serrate, pectinate or clubbed antennae all are apomorphic. In the case of the Ethiopian and Madagascan Agaristidae, however, genera with clubbed antennae (which form the vast majority of the genera considered here as Agaristid) form a rather homogenous group, with three easily recognizable subgroups, viz. the *Heraclia*-group, the *Aegoceropsis*-group, and the *Brephos*-group, all three with a basically similar pattern of markings. The "hererocerous" genera, on the other hand, lack homogeneity in that respect, except perhaps *Charitosemia* on the continent, and *Schausiana* in Madagascar. Of the genus *Aletopus* only the female is known, with filiform antennae, and it is quite possible that the male has clubbed antennae, the facies of the moth being very alike that of the *Apaegocera*-complex; the question is further complicated by the fact that both genera seem to belong to a mimetic group around some Geometric genera, and including the butterfly genus *Euphaedra*. A further development consists in the presence, in some genera, of a series of well differentiated, rather stiff spinae towards and around the apex of the valva. This peculiar, apomorphic character (also found in some Oriental genera) cannot, however, be considered as a synapomorphy, because it occurs in both main series of Agaristid genera: in the "heterocerous" series, the genera *Leucovis*, *Paratuerta* and *Hemituerta* **gen. nov.** have spinulated valvae, in the "rhopalocerous" series, this is the case with the genera *Misa*, *Hypotuerta* **gen. nov.**, *Metagarista*, *Weymeria*, *Massaga*, *Massagidia*, *Hespagarista* and *Syfanoidea.*

As far as the writer is able to judge, neither of the main groups just referred to is monophyletic on the Ethiopian faunistic niveau. The seemingly oldest genera, such as *Heraclia*, *Aegoceropsis* and *Mitrophrys*, are descended from different genera of the monophyletic Oriental complex of genera. The "rhopalocerous" group is monophyletic on a paleotropical niveau, and many students would probably find that this is sufficient to consider the Ethiopian branch alike as monophyletic.

The eight or nine "heterocerous" genera, however, cannot be considered as a monophyletic sister-group of the former, even on the above basis. The more plesiomorphic genera, the genus *Tuerta* of the authors, split in the present revision in several genera (*Tuerta*, *Pseudotuerta*, *Tuertella*, *Hemituerta*) on structural grounds, and *Pristoceraea*, certainly have not much in common, as evidenced by the very peculiar male genitalia of the latter genus. *Paratuerta*, with serrate, *Leucovis* with ciliate, and *Ovios* with bipectinate antennae, are in the same case. The fact that these last genera occur in East and South Africa procures perhaps a valuable indication, but our present knowledge of the Agaristidae is much too fragmentary, especially regarding the early stadia, to allow of a better substantiated arrangement.

The following species have not been included in the present revision, being considered by the writer as belonging to the sister-family of the Agaristidae, i.e. the Noctuidae:

Eudryas (?) *liturata* AURIVILLIUS, Entomol. Tidskr., Stockholm, **13**: 187, 1892.
Paratuerta abrupta ROTHSCHILD, Ann. Mag. nat. Hist., London, (9) **14**:315, 192?.
Paratuerta undulata BERIO, Bull. Soc. entomol. ital., Firenze, Genova, **102**:25–26; fig. 6, 1970.
BERIO (1970) places the genus *Paratuerta* with the Noctuid subfamily Amphipyrinae, but it belongs to the Agaristidae, and *P. undulata* should accordingly be placed in a new genus of its own.

Fam. **Agaristidae**
A. Ethiopian Genera

Key to the genera

The following artificial table should permit the identification of the Ethiopian Agaristid genera, although the presence of many very similar genera makes the identification sometimes slightly doubtful.

1	Antennae filiform, sometimes dilated towards the extremity	4
1′	Antennae ciliate, serrate or bipectinate	2
2	Antennae bipectinate	***Ovios***, p. 118
2′	Antennae ciliate or serrate	3
3	Antennae ciliate	***Leucovis***, p. 121
3′	Antennae serrate	***Paratuerta***, p. 104
4	Antennae not dilated towards the extremity	5
4′	Antennae dilated towards the extremity	11
5	Abdomen of male with a very long hairbrush	***Andrhippuris***, p. 4
5′	Abdomen of male without long hairbrush	6
6	Third segment of the palpi long	7
6′	Third segment of the palpi moderately long	8
7	Third segment of palpi very slender, naked	***Aletopus***, p. 10
7′	Third segment of palpi not very slender	***Charitosemia***, p. 116

8 Palpi not longer than the diameter of the eye 9
8′ Palpi longer than the diameter of the eye 10
9 Third segment of palpi short, bent . . . . ***Pseudotuerta*, p.** 113
9′ Third segment of palpi minute, pointed . . . . ***Tuertella*, p.** 114
10 Palpi twice as long as the diameter of the eye . . . . ***Hemituerta*, p.** 106
10′ Palpi shorter than twice the diameter of the eye . . . . ***Tuerta*, p.** 110
11 Frons shaped more or less snoutlike . . . . ***Choeropais*, p.** 97
11′ Frons not so shaped 12
12 Frons short beaklike; base of the abdomen with long lateral hairtufts in ♂ . . . . ***Paraegocera*, p.** 48
12′ These characters not present, at least not together 13
13 Fore wings very narrow; size small (length of fore wing 17 mm) . . . . . . ***Aethodes*, p.** 56
13′ Fore wings normal; size mostly larger 14
14 Fore tarsi with the first segment dilated in ♂ . . . . ***Weymeria*, p.** 8
14′ Fore tarsi with the first segment not dilated in ♂ 15
15 Hind tibiae with a long hairtuft 16
15′ Hind tibiae without long hairtuft 17
16 Palpi upturned; in fore wings, vein 10 originating from near the extremity of the areole . . . . ***Omphaloceps*, p.** 72
16′ Palpi porrect; in fore wings, vein 10 originating some distance from the extremity of the areole . . . . ***Agoma*, p.** 70
17 Hind wings with veins 3, 4 and/or 6, 7 stalked 18
17′ Hind wings with veins 3, 4 and 6, 7 from a point, or slightly separated 22
18 In fore wings, vein 6 arising from the areole; in hind wings, veins 3, 4 from a point . . . . ***Syfanoidea*, p.** 46
18′ In fore wings, vein 6 arising from the upper angle of cell, or below the angle; in hind wings, veins 3, 4 stalked 19
19 Palpi upturned, with the last segment long, slender; fore wings with vein 6 arising from below the cell-angle . . . . ***Crameria*, p.** 94
19′ Fore wings with vein 6 arising from the cell-angle 20
20 Hind tarsi with long hair on the first two segments in ♂ . . . . ***Metagarista*, p.** 49
20′ Hind tarsi without long hair on the first two segments in ♂ 21
21 Palpi long; last segment short, porrect . . . . ***Brephos*, p.** 90
21′ Palpi moderate, upturned; last segment minute . . . . ***Hypotuerta*, p.** 109
22 Fore wings with vein 5 well separated from veins 3, 4 23
22′ Fore wings with vein 5 approximated to veins 3, 4 24
23 Fore wings with veins 3, 4 approximated, and vein 5 from the lower cell-angle; vein 6 originating near the upper cell-angle . . . . ***Rothia*, 40 and p.** 126
23′ Fore wings with veins 3, 4 approximated, and vein 5 widely apart; vein 6 originating at the upper cell-angle; abdomen with basal haircrests . . . . ***Apaegocera*, p.** 6
24 Fore wings with vein 6 arising from the areole 25
24′ Fore wings with vein 6 arising at the upper cell-angle 26
25 Palpi very long, porrect, with the two last segments of equal length; fore wings with vein 6 arising from the extremity of the areole . . . . ***Paida*, p.** 86
25′ Palpi upturned; last segment short, with long hair; fore wings with vein 6 arising from the base of the areole . . . . ***Schausia*, p.** 82
26 Fore wings with vein 10 stalked with 9 of 8+9 27
26′ Fore wings with vein 10 arising from the areole, and not stalked with 9 or 8+9 28
27 Palpi with the last segment short in ♂, long in ♀; abdomen with long tufts of yellow hair at the base; fore wings with vein 10 stalked with veins 8+9 at the extremity of the areole; veins 3, 4, 5 well separated . . . . ***Massaga*, p.** 99
27′ Palpi upturned, with the last segment long, porrect; fore wings with vein 10 stalked with vein 9; veins 3, 4, 5 arising from a point . . . . ***Heraclia*, p.** 12
28 Palpi with the last segment long 29
28′ Palpi with the last segment short 32
29 Fore wings with veins 3, 4, 5 much approximated 30
29′ Fore wings with veins 3, 4, 5 well separated 31
30 Mid- and hind tibiae spined . . . . ***Aegoceropsis*, p.** 58
30′ Mid- and hind tibiae not spined . . . . ***Pentelia*, p.** 38
31 Frons with a bifid process . . . . ***Hoplarista*, p.** 85

31′ Frons without bifid process . . . . *Mitrophrys*, p. 64
32 Fore wings with veins 3, 4, 5 approximated **33**
32′ Fore wings with veins 3, 4, 5 well separataed **34**
33 Male with a stridulating organ . . . . *Misa*, p. 79
33′ Male without stridulating organ; abdomen with a long terminal tuft . *Lophonotidia*, p. 77
34 Areole minute or absent; vein 10 from its middle; hind wings with anal area clothed with very long hair; fore tibiae with a terminal spine . . . . *Pseudospiris*, p. 95
34′ Combination of characters different **35**
35 Mid- and hind tibiae spined **36**
35′ Mid- and hind tibiae not spined **37**
36 Abdomen with basal crest; fore wings with vein 10 from near the extremity of the areole . . . . *Polacanthopoda*, p. 52
36′ Abdomen with crests on the first and second segments; fore wings with vein 10 very proximal . . . . *Acantuerta*, p. 122
37 Last palpal segment robust . . . . *Chaetostephana*, p. 41
37′ Last palpal segment rather slender **38**
38 Head and thorax clothed with long hair; first three urites crested; termen of fore wings crenulate . . . . *Hespagarista*, p. 44
38′ Head and thorax clothed with short hair; fore wings with the termen not crenulate . . **39**
39 Fore femora covered with long and dense hair . . . . *Epischausia*, p. 75
39′ Fore femora not densely hairy **40**
40 Abdomen of ♂ with lateral tufts; hind wings with veins 3, 4 slightly separataed . . . . *Pseudopais*, p. 88
40′ Abdomen without lateral tufts; hind wings with veins 3, 4 from a point . *Massagidia*, p. 102

Genus Andrhippuris KARSCH

(fig. 1)

1895 *Andrhippuris* KARSCH, Entomol. Nachr.Bl., Wien, **21**:348.– Species typica: *Andrhippuris caudaequina* KARSCH (by original designation).

Diagnosis: Proboscis fully developed. Palpi obliquely porrect, with the second joint fringed with rough scales, and the third short, thickly scaled; frons broad, with a large truncate corneous prominence, flattened beneath, with a nearly reniform raised rim at the extremity. Antennae nearly filiform, without distinct terminal dilatation, even in male. Tibiae moderately fringed with hair. Abdomen without crests, but with a very long terminal tuft in male. Nervation: in fore wings, vein 2 given off from about three-quarters of cell; 3, 4, 5 approximated, equidistant; discocellulars faintly angled inwards; 6 from the upper cell-angle; areole rather long, narrow; 7 from just below the angle of the areole; 8+9 from the extremity of the areole, 10 from the upper side of the same, near the extremity. In hind wings, vein 2 from three-fifths of the cell; 3 and 4 from the cell-angle; discocellulars faintly incurved; 5 obsolescent, from middle of discocellulars; 6 and 7 very short stalked.

Male genitalia: Uncus short, with the base triangular, becoming extremely narrow, compressed, with a distinct terminal hook. Tubus analis rather broad, well sclerified. Vincula very broad, with a very broad fold bearing in the distal portion a few short hairs, but no distinct bunch of stiff hair. Valva elongate, rather narrow, with costa and sacculus parallel, the former faintly incurved after middle, the latter correspondingly outcurved; apex narrow, cut obliquely; fold of sacculus rather narrow, with the harpe absent. Aedeagus rather small, slightly longer than half of costa, slender, nearly straigt; fultura inferior broad; proximal margin straight; sides converging distally; terminal margin with a longer median and shorter lateral prongs; fultura

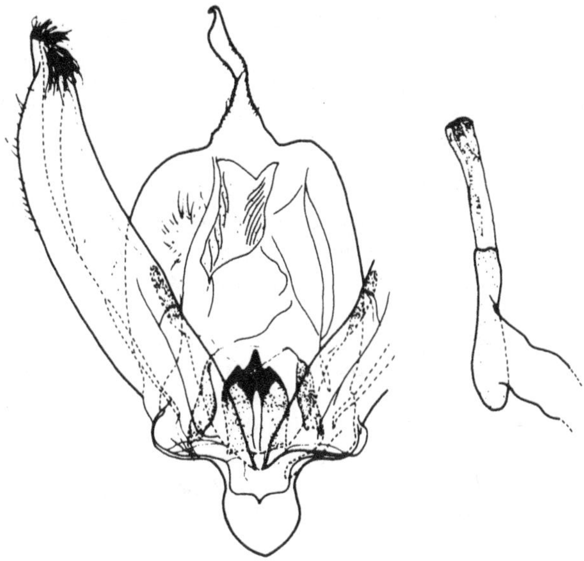

Fig. 1. *Andrhippuris caudequina* KARSCH. B. M. Noct. 8477

superior broad, with narrow lateral sclerifications. Saccus semi-elliptical, slightly broadening distally.

Distribution: West and Central Africa.

Andrhippuris caudaequina KARSCH

1895 *Andrhippuris caudaequina* KARSCH, Entomol. Nachr.Bl., Wien **21**:353; pl. 1, fig. 1 and 2. – Terra typica: Lower Guinea (Types, ♂, ♀, in Berlin Museum).

Diagnosis: Head and thorax black; first and second palpal joints, frons and sides of head, collar and shoulders spotted with white; thorax with some orange scales; hind tibiae fringed with yellow hair; tarsi slightly ringed with white, abdomen black with subdorsal orange patches forming dorsal bands on the first three urites, and a sublateral series of white patches. Fore wings rufous orange, mostly with a carmine suffusion; apical area suffused with golden brown; a basal black patch with two white striae on it, forming a fascia on dorsum, with patches of bluish-white scales on it just before the cell-end; a lunulate black discocellular spot with bluish-white lunules in centre; in male, a slight black suffusion at the lower cell-angle; before tornus, a black fascia with a few bluish-white scales; cilia blackbrown, spotted with white. Hind wings rufous orange, with a black terminal band with slight vein-streaks; cilia as on fore wings. Length of fore wing 21–28 mm.

Distribution: Lower Guinea; most of the Congo Basin: Kinshasa, Sankuru, Shaba, Zambia. There are no specimens from West Africa (except Kinshasa) in the British Museum nor at Tervuren.

Genus Apaegocera HAMPSON
(fig. 2)

1905 *Apaegocera* HAMPSON, Ann. Mag. nat. Hist., London, (7) **15**:449. – Species typica: *Apaegocera argyrogramma* HAMPSON (by original designation).

Diagnosis: Proboscis fully developed. Palpi upturned, the second joint fringed with long hair in front forming a pointed tuft at the extremity, the third joint long, porrect, dilated at the extremity. Antennae moderately dilated towards the extremity. Frons with a truncate conical process with a raised rim at the extremity. Tibiae nearly smooth, without spines. Abdomen with dorsal crests on the basal segments. Nervation: In fore wings, vein 2 given off slightly beyond middle of the cell; 3 and 4 approximated, distance between 4 and 5 nearly three times that between 3 and 4; upper discocellular incurved; 6 from the cell-angle; areole very narrow, 7, stalk of 8+9 and 10 from the extremity of the areole. In hind wings, vein 2 arising from about three quarters of the cell; 3, 4 and 6, 7 from the cell-angles; discocellulars very faintly incurved, 5 obsolescent, from middle of the discocellulars; 8 anastomosing with the cell shortly after base.

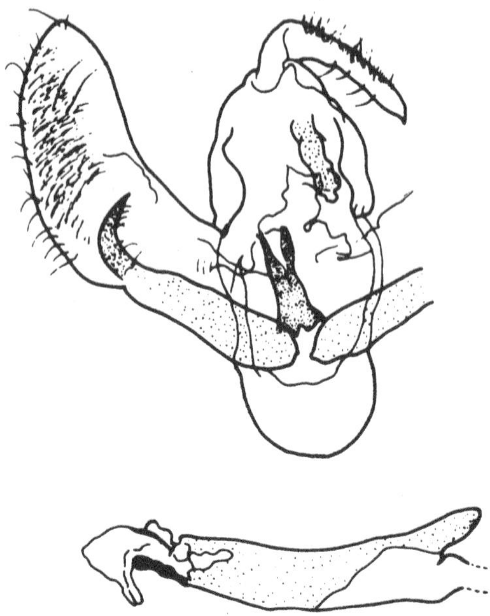

Fig. 2. *Apaegocera aurantiipennis* HAMPSON. B. M. Noct. 8579. Typus.

Male genitalia: Uncus about two-thirds of costa in length, compressed, arched at base, otherwise nearly straight, with a hardly marked terminal hook. Tubus analis short but broad, sclerified. Vinculum with the fold triangular, bearing s small bunch of hair. Valva broadly banana-shaped, with the costa on the whole faintly incurved, the apex broadly rounded, and the sacculus faintly incurved before the third, then outcurved; fold very broad basally, then narrowing, reaching to just before one-third of the sacculus, and ending in a small sickle-shaped harpe. Aedeagus very slightly shorter than the valva, moderately robust, somewhat more so proximally, faintly arched, bearing terminally a short stripe of cornuti; fultura inferior triangular, with the proximal margin notched at middle, and with a terminal rounded process. Saccus roughly semi-circular.

Distribution: West and Central Africa.

Key to the species

1 Fore wings reddish brown; hind wings orange . . . . ***A. argyrogramma*****, p. 7**
1′ Both pairs of wings orange . . . . ***A. aurantipennis*****, p. 7**

Apaegocera argyrogramma HAMPSON

1905 *Apaegocera argyrogramma* HAMPSON, Ann. Mag. nat. Hist., London, (7) **15**:449. – Terra typica: Obuassi, Ashanti (Type, ♂, in British Museum).

Diagnosis: Head and thorax black; extremities of the second and third palpal joints white, sides of frons white; vertex, collar, tegulae and thorax above spotted with white, tegulae edged with fulvous; abdomen black with white segmental lines; the three proximal urites fulvous with bluish-black dorsal crests; pectus and femora fulvous; tibiae and tarsi dark brown ringed with white. Fore wings fulvous, tinged with dark brown, extending on costa to the middle, and on dorsum to near the anal angle; distal area black; at the base of costa a small black and white patch; upper cell-angle with a small round white spot surrounded by a silvery-blue Y-shaped mark; subapically in the black area, an oblique band of five white spots in interspaces II to VII, the three upper spots conjoined. Hind wings orange red with a rather broad terminal black band, nearly uniform in width, slightly expanding at vein 2, and tapering to a point on the anal margin; cilia checquered black and white. Length of fore wing 29 mm.

Distribution: Ashanti; Ivory Coast.

Apaegocera aurantipennis HAMPSON

1912 *Apaegocera aurantipennis* HAMPSON, Ann. Mag. nat. Hist., London, (8) **10**:611. – Terra typica: Kampala, Uganda (Type, ♂, in British Museum).
1923 *Apaegocera joycei* HULSTAERT, Rev. Zool. Bot. Afr., Bruxelles, **11**:204. – Terra typica: Lake Albert, Zaire (Type, ♂, in Museum Tervuren).

Diagnosis: Head and thorax black; first and second palpal joints with white spots; frons white laterally; collar, tegulae, meso- and metathorax marked with white; extremities of tegulae and sides of metathorax with orange-red pilosity; pectus orange; femora streaked with orange; tibiae at the extremities and tarsi ringed with white, abdomen black with narrow white segmental bands; sides with some orange at the base. Wings reddish orange. Fore wings with the terminal area blackish brown, very broad at costa and extending as a triangular patch into the upper extremity of the cell where there is a white spot on it, preceded by a silvery blue dot, – and narrowing to the tornus, a white subapical band in the black area, between veins 4 and 7, traversed by dark streaks on veins 5 and 6, and small subterminal spots in interspaces II and II; a silvery blue discocellular bar; cilia with some white above the tornus. Hind wings with a narrow (1–1,5 mm wide) blackish brown terminal band, expanding slightly on vein 2 and on the anal margin; cilia chequered blackish brown and white. Length of fore wing 27–29 mm.

Distribution: Uganda; North-East Zaire (Stanleyville, Kibali-Ituri); Gabon (fide B. LAPORTE).

Genus Weymeria KARSCH
(fig. 3)

1895 *Weymeria* KARSCH, Entomol. Nachr.Bl., Wien, **21**:347. – Species typica: *Xanthospilopteryx athene* WEYMER (by monotypy).
1901 *Veymeria* HAMPSON (ex errore pro *Weymeria*), Cat. Lepidopt. Phal. Brit. Mus., **3**:548.

Diagnosis: Proboscis fully developed. Palpi oblique; first joint fringed with long hair; second joint nearly smoothly scaled; third joint long, slender, naked, por-

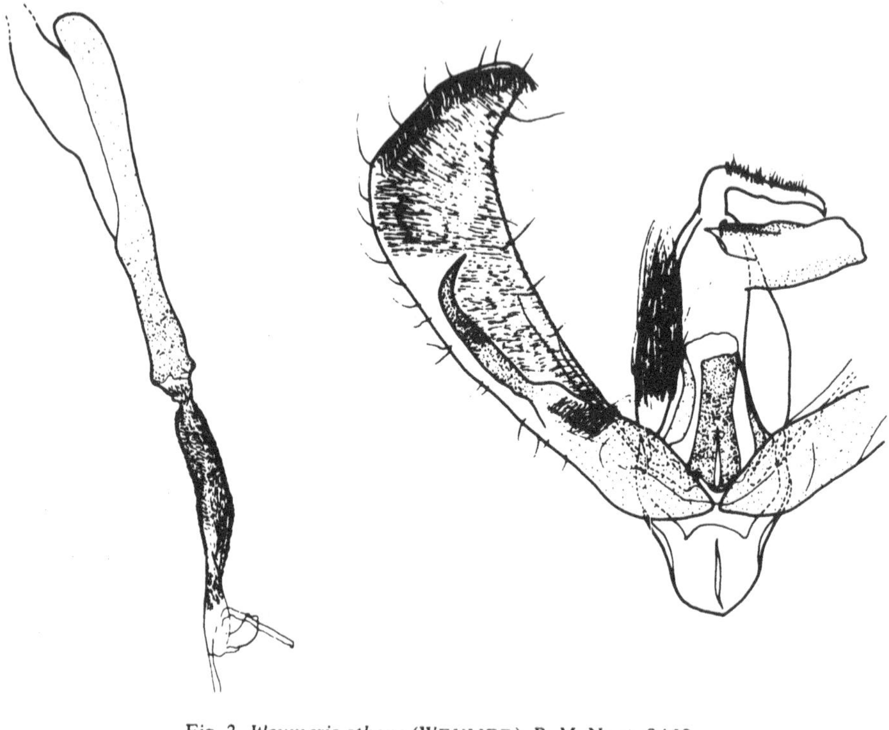

Fig. 3. *Weymeria athene* (WEYMER). B. M. Noct. 8459

rect. Antennae slightly dilated towards the extremity, less so in female. Frons with a truncate conical prominence, with raised rim at the extremity. Tibiae fringed with rough hair; fore tarsi in male with the first joint not much longer, but much more dilated than the second. Nervation: in fore wings, vein 2 arising from two-thirds of the cell; 3 and 4 separated; 5 given off from the lower fifth of the discocellulars which are faintly incurved; areole narrow; 6 from the lower cell-angle; 7 and the short stalk of 8+9 from the lower angle of the areole; 10 from the upper angle of the same. In hind wings, vein 2 from the seven-ninths of the cell; 3, 4 and 6, 7 from the cell-angles; discocellulars faintly incurved; 5 obsolescent, from the middle of the discocellulars, 8 fused with cell at base.

Male genitalia: Uncus about two-fifths of costa in length, compressed, bent at a right angle at about one-quarter of its length, then nearly straight, with a rather indistinct terminal hook. Tubus analis short and broad, sclerified both dorsally and ventrally. Vinculum narrow, with a strong bunch of stiff hair before the middle. Valva very elongate, rather narrow, slightly broadening distally, with the apical portion upturned; termen bearing a row of stiff, spicule-like hair; fold of sacculus very broad proximally, broader than the valve at base, then narrowing and ending in a slender, straight harpe, bent only terminally, reaching beyond the middle of the valva. Aedeagus about seven-tenths of costa in length, slender, nearly straight; everted vesica about two-thirds of aedeagus in length, covered, except apically, with minute spicules; two subterminal, nearly filiform processes; fultura inferior elongate and narrow, with a proximal projection at middle, and the sides faintly incurved; fultura superior broader, with the lateral margins sclerified. Saccus semi-elliptical.

Distribution: East Africa.

Weymeria athene (WEYMER)

1892 *Xanthospilopteryx athene* WEYMER, Stettin. entomol. Ztg., **53**:101. – Terra typica: Tanganyika (Type, in Berlin Museum).
1895 *Weymeria athene* – KARSCH, Entomol. Nachr. Bl., Wien, **21**:347; pl. 1, fig. 4.
1913 *Weymeria athene* aberr. *suffusa* JORDAN in SEITZ, Großschmetterl. Erde, **15**:11. – Terra typica: Tanganyika (Type, ♀, in British Museum).
1901 *Veymeria athene* – HAMPSON, Cat. Lepidopt. Phal. Brit. Mus., **3**:584; fig. 250.

Diagnosis: Head black, spotted with white and orange; palpi black with the first joint fringed with orange hair; thorax orange; collar spotted with black; pectus and legs black, coxae and spots on tibiae and tarsi white; abdomen orange with a series of black dorsal spots; last two urites and underside banded with black and white. Fore wings orange, more reddish at the base of dorsum; costa black; terminal area from mid-costa to tornus black, incurved in interspace I, with two large pale orange dots; cilia black tipped with white. Hind wings orange, more reddish basally; a terminal black band, about 6 mm wide in the subcostal region, and about $2^1/_2$ mm towards the anal angle, bearing four pale orange dots; cilia black, white near apex and the anal angle. Length of fore wing 25–28 mm.

Distribution Tanganyika; coast of Tanzania.

Genus Aletopus JORDAN
(fig. 4)

1926 *Aletopus* JORDAN, Novit. Zool., **33**:376. – Species typica *Aletopus imperialis* JORDAN (by original designation).

Diagnosis: Proboscis fully developed. Antennae of female not dilated towards the extremity. Frons with a conical truncate prominence, with raised rim at the extremity, the supra-oral ridge medially produced into a short sharp tooth. Palpi upturned, the second joint fringed with long hair in front, the third joint long, naked, somewhat porrect. Tibiae of female without long hair, and without spinae on mid and hind tibiae. Wings relatively short and rounded, particularly the hind wings. Neuration: in fore wings, veins 3, 4, 5 separated, vein 3 well before the cell-angle, being farther from 4 than this is from 5, 6 from the upper cell-angle; 7 and the long stalk of 8+9 from the extremity of the areole; 10 from the upper side of the latter, before the apex. In hind wings, vein 3 from well before the cell-angle; 5 from the middle of discocellulars; 6 and 7 stalked.

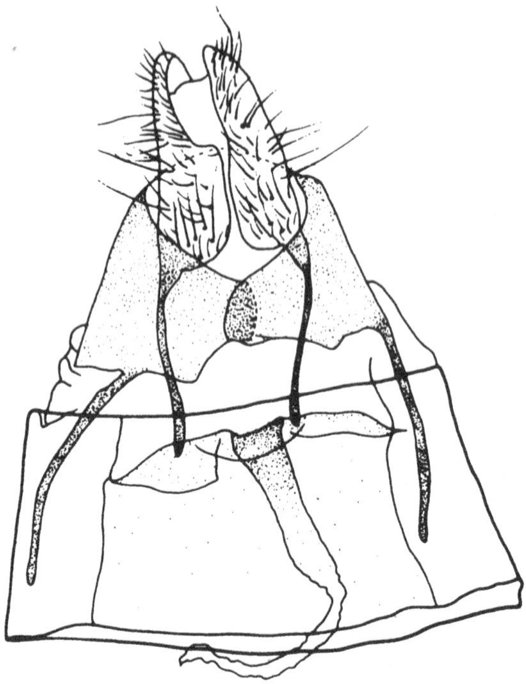

Fig. 4. *Aletopus imperialis* JORDAN B. M. Noct. 8298. Typus

Female genitalia: Papillae anales relatively very large; apophyses posteriores about as long as the papillae, and slightly shorter than the apophyses anteriores. Sterigma incurved at middle of the proximal margin. Ductus bursae sclerified, about as long as the apophyses posteriores. Bursa copulatrix small, without signum.

Distribution: Tanzania; Congo (Brazzaville)?

Aletopus imperialis JORDAN

1926 *Aletopsu imperialis* JORDAN, Novit. Zool. **33**:377. – Terra typica: Usambara, Tanganyika Territory (Type, ♀, in British Museum).

Diagnosis: Female: Body, antennae, palpi and legs black; basal joint of the antennae, extremities of the first and the second palpal joints, frons, head, pronotum, extremities of coxae, tibiae, and the first tarsal joints spotted with white; two white spots on each side of the abdominal segments, a white spot on the last segment, and a white ventral spot on the three proximal segments, the six first abdominal tergites with an orange segmental line, meso- and metanotum orange, black at middle. Fore wings rufous from the base to vein 2; rest of the wing black, with a white discal band from near costa across the discocellulars to beyond vein 2, broadening dorsally, distally, a round white spot on the second submedian vein; cilia white apically. Hind wings scarlet with a black terminal band bearing a row of five white dots, broadest from below the apex, cilia white apically. Length of fore wing 24 mm.

Remark: As long as there is but a single female known, the genus **Aletopus** must be maintained.

Distribution: Tanzania.

Species incertae sedis

Aletopus (?) ruspina (AURIVILLIUS)

1909 *Tuerta ruspina* AURIVILLIUS, Ark. Zool., Uppsala, **5**:9–10; fig. 43. – Terra typica: Bonga, "French Congo" (Type, ♂, in Brussels Museum; lost).
1913 *Schausia ruspina* – JORDAN in SEITZ, Großschmetterl. Erde, **15**:17.
1920 ? *Veymeria* (sic) *ruspina* HAMPSON, Cat. Lepidopt. Phal. Brit. Mus., Suppl. **2**:592 (species dubia).

Diagnosis: Black; palpi ringed with white; sides of frons and points at base of the antennae white, collar, shoulders and tegulae with double white points; tibiae narrowly white at the extremities; pectus and the first three urites with fulvous pilosity; abdomen with two series of white points; underside with white segmental lines. Fore wings orange; apical area from about two-thirds of costa to tornus black, slightly angled outwards on vein 2: an oblique subterminal series of white spots, small except the subcostal one; a narrow plumbeous streak on the discocellulars. Hind wings orange with a black terminal band about 3 mm wide, with a series of small white spots. Length of fore wing 29 mm.

Distribution: Bonga, "French Congo".

Remark: Careful investigations failed to locate the unique type in the collection of the Brussels Museum, and no similar specimens in other museums are known to the writer. The taxonomic position of the species must remain uncertain as long as no further specimens are available. It has been tentatively placed here with the genus *Aletopus* after comparison with the unique female of *A. imperialis.* The latter genus may turn out to be a synonym of *Weymeria* (cf. above). A puzzling feature is the uncommonly stout body of ? *A. ruspina* as figured by AURIVILLIUS, but since the unique type is unfortunately lost, there is no way to investigate the matter.

Genus **Heraclia** HÜBNER

(fig. 5)

1820 *Heraclia* HÜBNER, Verz. bek. Schmetterl., **12**:180. – Species typica: *Noctua geryon* FABRICIUS (by subsequent designation, HAMPSON 1920).

1858 *Xanthospilopteryx* WALLENGREN, Oefv. vet. Acad. Förh., **15**:83. – Species typica: *Xanthospilopteryx geryon* – WALLENGREN (nec FABRICIUS). = *Eusemia superba* BUTLER (by original designation).

Diagnosis: Proboscis fully developed. Antennae distinctly dilated towards the extremity. Palpi with the second joint obliquely upturned and fringed with long hair in front, and the third joint long, porrect. Frons not narrowed behind, with a slight truncate conical projection, with a raised rim at the extremity. Mid and hind tibiae slightly fringed with rough hair above, and sometimes beneath. Nervation: In fore wings, vein 2 given off at three-quarters of cell; 3, 4, 5 from the lower cell-angle; discocellulars angled inwards; 6 from the upper cell-angle; areole narrow; 7, 8+9, 10 from the extremity thereof, 10 mostly from 9. In hind wings, vein 2 given off at two-thirds of cell; 3, 4 and 6, 7 from the cell-angles; discocellulars incurved; 5 obsolescent, from middle of discocellulars.

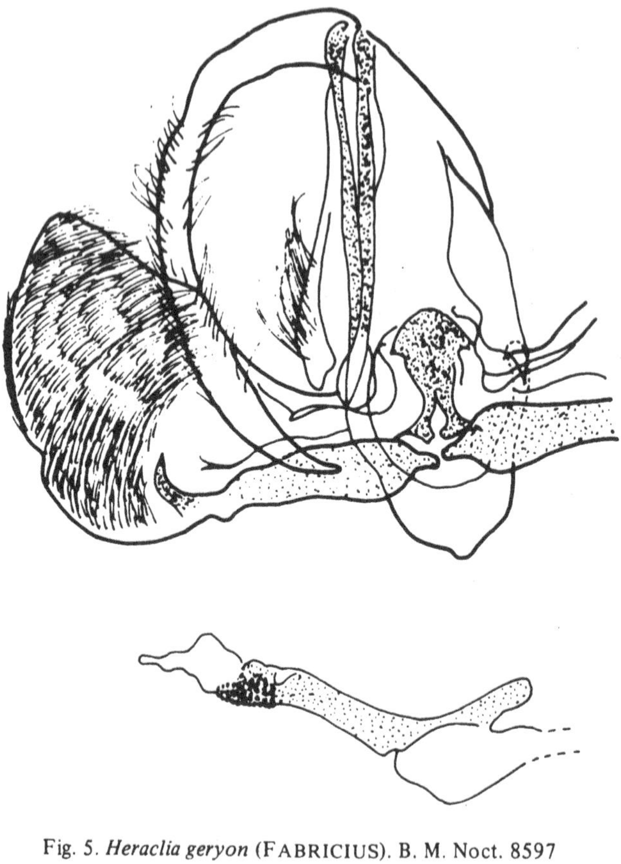

Fig. 5. *Heraclia geryon* (FABRICIUS). B. M. Noct. 8597

Male genitalia: Uncus generally as long as, or longer than valva, but distinctly shorter in some species, narrow, arched, with a terminal hook. Tubus analis typically reaching the base of the valvae, often spatulate terminally, but much shorter in a number of species. Valva roughly triangular, with the apex more or less pointed, and the sacculus broadly outcurved; harpe relatively small, hooked, but in some species straight and directed costad. Aedeagus rarely robust, mostly distinctly shorter than the valva, very often bearing terminally patches of minute cornuti, fultura inferior very variable, in the type species mushroom-shaped. Saccus very short, rounded or slightly angulate.

Distribution: West, South, Central and East Africa.

Key to the species

1 Fore wings with the ground colour fulvous orange 2
1′ Fore wings with the ground colour black or yellow 3
2 Fore wings with white spots in the black apical area . . . . . *H. poggei*, p. 35
2′ Fore wings without white spots in the black apical area . . . . . *H. catori*, p. 36
3 Hind wings with a black discocellular spot 4
3′ Hind wings without black discocellular spot 5
4 Black discocellular spot very large . . . . . *H. mozambica*, p. 37
4′ Black discocellular spot small . . . . . *H. viettei*, p. 37
5 Abdomen banded black and yellow or orange 10
5′ Abdomen not banded black and yellow or orange 6
6 Abdomen with the ground colour yellow or orange 7
6′ Abdomen black 33
7 Abdomen yellow . . . . . *H. flavipennis*, p. 30
7′ Abdomen yellow or orange tinged with dark or reddish 8
8 Abdomen tinged with blackish 27
8′ Abdomen tinged with pink or reddish 9
9 Abdomen at base with two black dorsal spots . . . . . *H. thruppi*, p. 24
9′ Abdomen at base with a yellow tuft of hair . . . . . *H. durbanica*, p. 25
10 Abdomen on underside banded black and white or yellow 11
10′ Abdomen on underside black spotted white 22
11 Abdomen with all ventral markings white or yellowish white 12
11′ Abdomen with the ventral markings at least partly yellow or orange 18
12 Abdomen with pale yellow segmental lines . . . . . *H. contigua*, p. 23
12′ Abdomen with white ventral bands 13
13 Collar spotted with yellow or orange . . . . . *H. geryon*, p. 14
13′ Collar spotted with white 14
14 Hind wings with the costal edge black in distal part only 15
14′ Hind wings with the costal edge black to base 16
15 Abdomen tipped with yellow; fore wings with the dorsal streak present . *H. pallida*, p. 20
15′ Abdomen tipped with black; fore wings with the dorsal streak absent . *H. deficiens*, p. 22
16 Hind wings white with a broad black margin . . . . . *H. medeba*, p. 23
16′ Hind wings yellow, sometimes tinged with blackish 17
17 Hind wings yellow with a narrow black margin . . . . . *H. terminatis*, p. 21
17′ Hind wings yellow, very strongly tinged with blackish . . . . . *H. atrifusa*, p. 22
18 Abdomen on underside with orange bands 19
18′ Abdomen on underside with orange bands except on the last three urites where the bands are white . . . . . *H. africana*, p. 19
19 Abdomen with the bands on underside very indistinct . . . . . *H. aisha*, p. 24
19′ Abdomen with the bands on underside broad 20
20 Anal tuft black in both sexes; fore wings with a continuous yellow medial band . . . . *H. xanthopyga*, p. 21
20′ Anal tuft black in ♂, yellow in ♀; fore wings with a series of two medial spots 21
21 Fore wings with the subapical band oval; hind wings carmine-red . . . . . *H. perdix*, p. 19
21′ Fore wings with the subapical band narrow; the lower spot short and nearly isolated; hind wings salmon pink . . . . . *H. flavisignata*, p. 20

22 Abdomen on underside with three series white spots **23**
22′ Abdomen on underside with one or two series white spots **24**
23 Hind wings with the black terminal band broad towards the anal angle, and with an orange spot on underside below vein 2 . . . . ***H. pardalina***, p. 15
23′ Hind wings with the black terminal band narrow towards the anal angle, and without orange spot on underside below vein 2 . . . . ***H. superba***, p. 17
24 Abdomen on underside with a sublateral series of white or yellowish dots **25**
24′ Abdomen on underside with a medial series of white dots **26**
25 Tegulae spotted with yellowish; tibiae banded with yellowish; tarsi spotted with white; cilia on hind wings white apically . . . . ***H. atriventralis***, p. 15
25′ Tegulae spotted with white; tibiae mostly orange; tarsi ringed yellowish; cilia on hind wings orange apically . . . . ***H. butleri***, p. 16
26 Abdomen on upperside banded yellow and black . . . . ***H. jugans***, p. 16
26′ Abdomen on upperside yellow . . . . ***H. limbomaculata***, p. 36
27 Fore wings with the ground colour yellow . . . . ***H. zenkeri***, p. 27
27′ Fore wings with the ground colour black **28**
28 Hind wings with the costal edge black to base . . . . ***H. longipennis***, p. 31
28′ Hind wings with the costal edge black in distal part only **29**
29 Collar and tegulae spotted with yellow . . . . ***H. monslunensis***, p. 33
29′ Collar and tegulae spotted with white **30**
30 Hind wings with the basal area tinged with black **31**
30′ Hind wings with the basal area not tinged with black **32**
31 Fore wings spotted with yellow . . . . ***H. karschi***, p. 28
31′ Fore wings spotted with white . . . . ***H. gruenbergi***, p. 29
32 Fore wings with the subapical band continuous . . . . ***H. aurea***, p. 35
32′ Fore wings with the subapical band broken into spots . . . . ***H. aemulatrix***, p. 31
33 Fore wings with the markings whitish . . . . ***H. hornimani***, p. 25
33′ Fore wings with the markings yellow **34**
34 Abdomen with the anal tuft orange; fore wings rather narrow and elongate **35**
34′ Abdomen with the anal tuft black, with some orange admixture **36**
35 Fore wings with a subapical series of spots . . . . ***H. hypercompoides***, p. 26
35′ Fore wings with a continuous subapical band . . . . ***H. elongata***, p. 27
36 Hind wings red or orange above **37**
36′ Hind wings black above **38**
37 Fore wings with the spots yellow, moderate; hind wings with a black terminal band . . ***H. nigridorsa***, p. 33
37′ Fore wings with the spots pure white, extensive; hind wings red with a black margin all around . . . . ***H. kivuensis***, p. 33
38 Fore wings with the medial and subapical series of spots well separated and broken up by black veins . . . . ***H. buchholzi***, p. 34
38′ Fore wings with the medial and subapical series of spots fused into a large square patch, with a large black discocellular mark within . . . . ***H. annulata***, p. 35

Heraclia geryon (FABRICIUS)

1781 *Noctua geryon* FABRICIUS, Spec. Ins., **2**:216. – Terra typica: Africa (Type not designated).
1782 *Noctua euphemia* STOLL, Uitheemsche Kapellen, **4**:pl. 345 A. – Terra typica: Africa (Type not designated).
1820 *Heraclia geryon* – HÜBNER, Verz. bek. Schmetterl., **12**:180.
1854 *Eusemia euphemia* – WALKER, List Lepidopt. Het. Brit. Mus., **1**:50.
1881 *Eusemia nugatrix* WESTWOOD, Oates' Matabeleland:356. – Terra typica: Cape Palmas (Type, ♂, in Museum Oxford University).
1891 *Xanthospilopteryx geryon* – KIRBY, Trans. R. entomol. Soc. London, **1891**:287–288.
1892 *Xanthospilopteryx euphemia* – KIRBY, Syn. Cat. Lepidopt. Het.:29.
1892 *Xanthospilopteryx nugatrix* – KIRBY, Syn. Cat. Lepidopt. Het.:29.
1913 *Xanthospilopteryx geryon* forma *flava* JORDAN in SEITZ, Großschmetterl. Erde, **15**:6. – Terra typica: Not designated.

D i a g n o s i s : Head and thorax black; palpi, sides of frons, back of head, tegulae with white points, collar broadly spotted with orange; chin, patches on pectus and coxae, streaks on femora and rings on tarsi white; tibiae spotted with orange, abdomen black, with orange dorsal bands and white lateral spots and ventral bands, anal tuft black in ♂, orange in ♀. Fore wings black, marked with orange, viz.: two basal spots, a subbasal triangular cell-spot, sometimes with streaks above and below it; a dorsal streak; square patches in cell-end and below base of vein 2, latter mostly larger; an oblique postmedial band from below costa to vein 2, sometimes connected with a patch below vein 2; a spot above tornus; some silvery-blue scales in a below cell near base, in and below middle of cell, and on the discocellulars, cilia white apically. Hind wings carmine-red or orange-red (aberr. *flava*), with black costa and a broad terminal band; cilia white apically. Length of fore wing 32–34 mm.

D i s t r i b u t i o n : West and Central Africa to Upper Egypt and Burundi.

Heraclia pardalina (WALKER)

1869 *Eusemia pardalina* WALKER, Proc. nat. Hist. Soc. Glasgow, **1**:329; pl 5, fig. 1 – T e r r a t y p i c a : Congo (Type, ♀, in British Museum).
1875 *Eusemia ochracea* BUTLER, Ann. Mag. nat. Hist., London, (4) **15**:142 – T e r r a t y p i c a : Congo (Type, ♀, in British Museum).
1891 *Xanthospilopteryx pardalina* – KIRBY, Trans. v. entomol. Soc. London, **1891**:290.
1892 *Xanthospilopteryx ochracea* – KIRBY, Syn. Cat. Lepidopt. Het.:29.
1893 *Xanthospilopteryx kirbyi* CARPENTER, Proc. r. Dublin Soc., **8**:308. – T e r r a t y p i c a : Not designated.

D i a g n o s i s : Head and thorax black; sides of palpi, chin, coxae, femora and tibiae at the extremities, and tarsal joints spotted with white; sides of frons, vertex, collar and tegulae spotted with ochreous yellow; tibiae spotted with orange; abdomen black, banded orange above, and with three ventral series of ochreous-white spots; anal tuft orange in ♀. Fore wings black, spotted with orange yellow, viz., about six points in basal area; an antemedial spot from just above to just below cell; a small medial spot above vein 1; an elongate spot below vein 1; a square spot in cell-end, and another below base of vein 2; an oblique postmedial band between veins 3 and 8; spot above tornus; silvery-blue marks in a below cell, on the discocellulars, and on veins 1 and 2; cilia yellowish apically. Hind wings orange with some black at the base; a broad terminal black band, running on costa inwards to base; cilia deep yellow apically. Length of fore wing 27 mm.

D i s t r i b u t i o n : Zaire, Angola.

Heraclia atriventralis (HAMPSON)

1910 *Xanthospilopteryx atriventralis* HAMPSON, Proc. zool. Soc. London, **1910**:396; pl. 36, fig. 13. – T e r r a t y p i c a : Serenje District, Northeast Rhodesia = Zambia (Type, ♂, in British Museum).
1920 *Heraclia atriventralis* – HAMPSON, Cat. Lepidopt. Phal. Brit. Mus., Suppl. **2**:551.

D i a g n o s i s : Head and thorax black; the first and the second palpal joints, sides of frons, vertex, and collar spotted with white; tegulae spotted with yellowish;

joints of femora and tibiae, and tarsi with white points; tibiae banded yellosiwh, abdomen black with dorsal and lateral yellow bands, underside black with a sublateral series of whitish or yellowish dots. Fore wings black, spotted with yellow, wiz., a subbasal spot below costa; an elongate antemedial spot below costa, a small spot above vein 5; a spot in cell-middle; a medial spot above vein 1; an oblique band from below middle of costa to vein 1 towards the tornus, ending in a point; an oblique subapical band from below costa to vein 3 near the termen; a small subterminal spot above vein 1; cilia white apically. Hind wings carmine-red with a black terminal band, broader towards the apex; cilia white apically. Length of fore wing 25 mm.

Distribution: Zambia.

Heraclia jugans (JORDAN)

1813 *Xanthospilopteryx jugans* JORDAN in SEITZ, Großschmetterl. Erde, **15**:7; pl. 4 h. – Terra typica: Sangui Valley, Lake Nyasa (Type, ♂, in British Museum).
1920 *Heraclia jugans* – HAMPSON, Cat. Lepidopt. Phal. Brit. Mus., Suppl. **2**:551.

Diagnosis: Head and thorax black; the first and the second palpal joints, sides of frons, vertex, collar, mesothorax and pectus spotted with white; tegulae with yellowish white spots; cheeks white; fore coxae and femora with some white; tibiae broadly yellowish, extremities of tibiae and tarsi ringed with white, abdomen black with yellow rings above, and a series of white points below; anal tuft black in ♂, black then broadly yellow, then again black, and deep orange at the extremity in ♀. Fore wings black; base with three white spots below costa, markings orange yellow, viz., two pairs antemedial spots in and below cell, the distal pair sometimes fused; an oblique medial band from costa to vein 1; an oblique postmedial band from below costa to vein 3 near the termen; a small spot above tornus, some silvery-blue at base, a bar in cell-middle, a discocellular streak, and postmedial spots on veins 1 and 2; cilia yellowish apically. Hind wings carmine-red, with a terminal blackish band, widening costally; cilia yellowish apically. Length of fore wing 29 mm.

Distribution: Malawi; Angola.

Heraclia butleri (WALKER)

1869 *Eusemia butleri* WALKER, Char. Lepidopt. Het.:111. – Terra typica: Natal (Type, in British Museum).
1881 *Eusemia adulatrix* WESTWOOD, Oates' Matabeleland:355; pl. G, fig. 1; pl. H, figs. 3, 3a and 3b. – Terra typica: Matabeleland, Rhodesia (Type, ♂, in Museum Oxford University).
1891 *Xanthospilopteryx butleri* – KIRBY, Trans r. entomol. Soc. London, **1891**:291; pl. 15, fig. 4.
1891 *Xanthospilopteryx adulatrix* – KIRBY, Trans. r. entomol. Soc. London, **1891**:290.
1901 *Rothia butleri* – HAMPSON, Cat. Lepidopt. Phal. Brit. Mus., **3**:576; pl. 53, fig. 13.
1901 *Rothia butleri* ssp. *adulatrix* – HAMPSON, Cat. Lepidopt. Phal. Brit. Mus., **3**:577.
1913 *Xanthospilopteryx butleri* forma *adulatrix* – JORDAN in SEITZ, Großschmetterl. Erde, **15**:7.
1913 *Xanthospilopteryx butleri* forma *completa* JORDAN in SEITZ, Großschmetterl. Erde, **15**:7. – Terra typica: not designated (Type, ♂, in British Museum).
1937/39 *Heraclia butleri* forma *adulatrix* – JANSE, Moths South Africa, **3**:20.

D i a g n o s i s : Body and fore wings black; sides of palpi and of frons, vertex, collar, tegulae and thorax above spotted with white; chin yellow; coxae and a great part of tibiae orange; tarsi ringed with yellowish, abdomen with dorsal orange bands and a sublateral series often obsolescent white spots; anal tuft orange at the extremity. Fore wings marked with yellowish, viz., two points below base of costa, and two below cell, an antemedial cell-spot; a slight spot above vein 1; a square spot at the cell-end, conjoined to a spot below cell; an oblique band beyond cell between veins 3 and 8, more or less excised on the media (hardly so in forma *adulatrix*); a spot above tornus; some silvery-blue marks, viz., in and below cell near base, in and below middle of cell; above vein 1; a discocellular bar; spots on veins 1 and 2; cilia yellow apically. Hind wings carmine-red with a black terminal band, angled inwards on vein 2, running inwards on costa short of middle; cilia orange apically in ♂. Length of fore wing 23 mm in ♂, 27 mm in ♀.

L a r v a l s t a d i a : Larva grey blue with fine black transversal stripes and spots on each segment; a reddish lateral line, head, legs and first and eleventh segments dorsally reddish-yellow; a few sparse hairs; prolegs blue grey; last pair reddish yellow. On *Vitis.* Pupa reddish brown, rugged in appearance.

D i s t r i b u t i o n : Souteast and East Africa (Natal, Transvaal, Mashonaland, Malawi, Rhodesia, Zambia, Tanzania).

Heraclia superba (BUTLER)

1875 *Eusemia superba* BUTLER, Ann. Mag. nat. Hist., London, (4) **15**:141; pl. 13, fig. 3. – T e r r a t y p i c a : Zululand (Type, ♂, in British Museum).
1891 *Xanthospilopterys superba* – KIRBY, Trans. r. entomol. Soc. London, **1891**:289.
1892 *Eusemia interniplaga* MABILLE, Bull. Soc. entomol. France, Paris, **1891**:184. – T e r r a t y p i c a : West Africa (Type, ♀, in British Museum).
1892 *Xanthospilopteryx interniplaga* – KIRBY, Syn. Cat. Lepidopt. Het.:899.
1897 *Xanthospilopteryx smithii* HOLLAND, in Donaldson's Smith Unknown African Countries: 408; fig. 3. – T e r r a t y p i c a : Dabulli (Type, ♂, In Carnegie Museum).
1920 *Heraclia superba* – HAMPSON, Cat. Lepidopt. Phal. Brit. Mus., Suppl. **2**:550.
1923 *Heraclia littera* HULSTAERT, Rev. Zool. Bot. Afr., Bruxelles, **11**:202. – T e r r a t y p i c a : "Congo Belge" = Zaire (Type, ♀, in Museum Tervuren).
1923 *Heraclia albocincta* HULSTAERT, Rev. Zool. Bot. Afr., Bruxelles, **11**:202. – T e r r a t y p i c a : "Congo Belge" = Zaire (Type, ♀, in Museum Tervuren).

D i a g n o s i s : Head and thorax black; sides of palpi and of frons, vertex, collar, tegulae spotted with white; fore coxae and extremities of femora white; tarsal joints with white points; tibiae spotted with orange; abdomen black, with dorsal orange bands; underside with three series white points; anal tuft black in ♂, yellow in ♀. Fore wings black; several bluish-white basal points; markings orange, viz., a small spot below cell; another avove vein 1, mostly reaching the dorsum, and followed by a large antemedial spot extending from above to below cell; a square spot below base of vein 2; a broad oblique postmedial band between veins 3 and 8; a spot above tornus; silvery-blue scales near base, and silvery-blue markings, viz., an oblique bar in middle of cell, with a streak above it and a spot in the submedial fold, a discocellular bar, and postmedial spots on veins 1 and 2; cilia white apically. Hind wings orange to carmine-red, with the costa and a broad terminal band black, broader apically; cilia white apically. Length of fore wing 25–38 mm.

D i s t r i b u t i o n : South, Central and East Africa.

Heraclia superba superba (BUTLER)

1913 *Xanthospilopteryx superba* ssp. *superba* – JORDAN in SEITZ, Großschmetterl. Erde, **15**:7; pl. 1f.

Diagnosis: Subapical band on fore wings broad, hind wings carminered.

Distribution: Natal, Transvaal, Mashonaland, Malawi, Rhodesia, Zambia, Burundi, South Zaire to North Kivu.

Heraclia superba ugandana JORDAN

1913 *Xanthospilopteryx superba* ssp. *ugandana* JORDAN in SEITZ, Großschmetterl. Erde, **15**:7. – Terra typica: Uganda (Type, ♂, in British Museum).

Diagnosis: Subapical band on fore wings narrower; spot in cell and square spot below base of vein 2 nearly always fused; hind wings orange-red.

Distribution: Uganda, Rwanda, East Zaire: Tanganyika, Kivu.

Heraclia superba basalifasciata (BARTEL)

1903 *Xanthospilopteryx basalifasciata* BARTEL, Verh. zool.-bot. Ges. Wien, **53**:120. – Terra typica: Central Africa.

1910 *Xanthospilopteryx tanganyikana* OBERTHÜR, Ann. Soc. entomol. France, Paris, **1910**:471; pl. 16, fig. 1. – Terra typica: M'Pala, Tanganyika (Type, ♀, in British Museum).

1911 *Xanthospilopteryx maculipes* STRAND, Mitt. zool. Mus. Berlin, **5**:301; fig. – Terra typica: Tanganyika (Type, ♀, in Berlin Museum).

1912 *Xanthospilopteryx superba* aberr. *rufoides* STRAND, Lepidopt. Cat., **5**:18. – Terra typica: Tanganyika (Type, in British Museum).

1913 *Xanthospilopteryx superba* ssp. *basalifasciata* – JORDAN in SEITZ, Großschmetterl. Erde, **15**:7.

1913 *Xanthospilopteryx superba basalifasciata* aberr. *separata* JORDAN in SEITZ, Großschmetterl. Erde, **15**:7.

Diagnosis: Fore wings with the subbasal spots fused (*basalifasciata*) or separated (*separata*), subbasal and medial bands more or less fused (*tanganyikana*), subbasal, medial and postmedial spots united into three bands (*maculipes*); hind wings orange.

Distribution: Tanzania; East Zaire: Kivu.

Heraclia superba incongruens (BUTLER), comb. nov.

1878 *Eusemia incongruens* BUTLER, Proc. zool. Soc. London, **1878**:38. – Terra typica: Atbara River, Ethiopia (Types, ♂, ♀, in British Museum).

1891 *Xanthospilopteryx incongruens* – KIRBY, Trans. r. entomol. Soc. London, **1891**:289; pl. 15, figs. 6 and 7.

1912 *Xanthospilopteryx superba* aberr. *incongruens* – STRAND, Lepidopt. Cat., **5**:18.

D i a g n o s i s : Fore wings with the subbasal spot above vein 1 small, not reaching the dorsum; postmedial band shorter (6 mm v. 8–9 mm); hind wings orange, sometimes tinged with red.

D i s t r i b u t i o n : Ethiopia.

Heraclia perdix (DRUCE)

1887 *Eusemia perdix* DRUCE, Proc. zool. Soc. London, **1887**:668. – T e r r a t y p i c a : Dalagoa Bay (Types, ♂, ♀, in British Museum).
1890 *Eusemia eoa* MABILLE, Bull. Soc. entomol. France, Paris, (6) **10**:123. – T e r r a t y p i c a : Afrique orientale (Type, ♀, in British Museum).
1891 *Xanthospilopteryx perdix* – KIRBY, Trans. r. entomol. Soc. London, **1891**:285.
1891 *Eusemia indecisa* BUTLER, Ann. Mag. nat. Hist., London, (6) **7**:50. – T e r r a t y p i c a : Central Africa (Type, ♂, in British Museum).
1892 *Xanthospilopteryx eoa* – KIRBY, Syn. Cat. Lepidopt. Het.:898.
1892 *Xanthospilopteryx indecisa* – KIRBY, Syn. Cat. Lepidopt. Het.:29.
1912 *Xanthospilopteryx perdix* aberr. *angustella* STRAND, Lepidopt. Cat., **5**:17. – T e r r a t y p i c a : not designated.
1913 *Xanthospilopteryx perdix* aberr. *augustella* JORDAN in SEITZ (ex errore pro *angustella*), Großschmetterl. Erde, **15**:5.

D i a g n o s i s : Body black; sides of palpi and of frons, vertex, collar, tegulae and thorax above with yellowish white points, pectus, coxae, femora, tibiae and tarsi marked with white; tibiae banded orange; abdomen with orange bands and anal tuft. Fore wings black; two yellowish basal points and a short streak above vein 1; markings orange, viz., an oblique sinuous antemedial band from the subcostalis to vein 1, a broad bar across cell; a square patch below base of vein 2; a broad band beyond cell between veins 3 and 8; a spot above tornus; some silvery-blue markings, viz., a cell-bar with spot below it; a discocellular bar; short postmedial streaks on veins 1 and 2, cilia white apically. Hind wings deep red; a broad black terminal band extending on costa before the middle, a red spot on termen below vein 2, sometimes absent (*indecisa*), cilia yellowish apically. Length of fore wing 32 mm.

D i s t r i b u t i o n : Natal, Mozambique, Malawi, Zambia, Tanzania.

Heraclia africana (BUTLER)

1875 *Eusemia africana* BUTLER, Ann. Mag. nat. Hist., London, (4) **15**: 142. – T e r r a t y p i c a : Natal (Types, ♂, ♀, in British Museum).
1881 *Eusemia meretrix* WESTWOOD, Oates' Matabeleland:355. – T e r r a t y p i c a : Matabeleland (Type, ♀, in Museum Oxford University).
1890 *Eusemia euphemia* – MABILLE (nec STOLL 1782), Bull. Soc. entomol. France, Paris, (6) **10**:124.
1891 *Xanthospilopteryx africana* – KIRBY, Trans. r. Soc. entomol. London, **1891**:287.
1891 *Xanthospilopteryx fatima* KIRBY, Trans. r. entomol. Soc. London, **1891**:288; pl. 15, fig. 2. – T e r r a t y p i c a : Tanzania: Taita (Type, ♂, in British Museum).
1892 *Xanthospilopteryx meretrix* – KIRBY, Syn. Cat. Lepidopt. Het.:29.
1913 *Xanthospilopteryx africana* forma *flava* JORDAN in SEITZ, Großschmetterl. Erde, **15**:5, pl. 1d. – T e r r a t y p i c a : not designated.
1920 *Heraclia africana* – HAMPSON, Cat. Lepidopt. Phal. Brit. Mus., Suppl. **2**:550.

D i a g n o s i s : Head and thorax black; sides of palpi and of frons, vertex, collar, tegulae, metathorax above, pectus and fore coxae spotted with white; tibiae spotted with orange; tarsi ringed with white; abdomen banded black and orange, underside with the distal bands white. Fore wings black; four basal bluish white spots; markings orange, viz., an oblique antemedial band from the submedian vein to vein 1, an oblique band across the cell-end, a square patch below base of vein 2; an oblique postmedial band to vein 3 towards termen; a spot above tornus, some silvery-blue markings, viz., an oblique spot in cell-middle, a few scales below cell, and a discocellular lunule; cilia white apically. Hind wings crimson-red or yellow (forma *flava*); a broad terminal black band, expanding on costa to before middle, cilia white apically. Length of fore wing 31–35 mm.

D i s t r i b u t i o n : Republic of South Africa: Cape Province, Namibia, Transvaal, Natal; Malawi; Tanzania; Uganda.

Heraclia flavisignata (HAMPSON)

1912 *Xanthospilopteryx flavisignata* HAMPSON, Ann. Mag. nat. Hist., London, (8) **10**:607. – T e r r a t y p i c a : Lusinga, Victoria Nyanza, British East Africa = Tanzania (Type, ♂, in British Museum).

1920 *Heraclia flavisignata* HAMPSON, Cat. Lepidopt. Phal. Brit. Mus., Suppl. **2**:549, pl. 70, fig. 18.

D i a g n o s i s : Head and thorax black; the two first palpal joints, sides of frons, vertex, collar, tegulae, meso- and metathorax with yellow points; pectus and femora with some yellow; tibiae striped with orange; tarsi ringed with yellow; abdomen banded orange and black above and below. Fore wings black; markings yellow, viz., four small basal dots; a small triangular antemedial spot in cell; two small oblique spots above vein 1; a broad oblique bar across the cell-extremity; an elliptical patch below cell; an oblique band from below costa postmedially to vein 4, separated by a black streak on vein 4 from a yellow spot below vein 4; a small spot above tornus; silvery-blue markings, viz., in basal area, an oblique line in middle of cell, a line on the discocellulars; cilia yellowish-white apically. Hind wings reddish-orange with a black terminal band, broad at apex and narrowing to tornus; cilia yellowish-white apically. Length of fore wing 30 mm.

D i s t r i b u t i o n : Luzinga, Tanzania.

Heraclia pallida (WALKER)

1854 *Eusemia pallida* WALKER, List Lepidopt. Het. Brit. Mus., **1**:52. – T e r r a t y p i c a : Not designated (Type, ♀, in British Museum).

1881 *Eusemia niveosparsa* WESTWOOD, Oates' Matabeleland:355. – T e r r a t y p i c a : Cameroon ? (Type, ♀, in Museum Oxford University).

1891 *Xanthospilopteryx niveosparsa* – KIRBY, Trans. r. entomol. Soc. London, **1891**:284.

1891 *Xanthospilopteryx pallida* – KIRBY, Trans. r. entomol. Soc. London, **1891**:286.

D i a g n o s i s : Head and thorax black; sides of palpi and of frons, occipiut, collar, tegulae with white points; stripes on coxae and on femora, rings on tarsi white; tibiae spotted with orange; abdomen black, ringed with yellow above, very broadly in ♂ on the proximal urites; last urites banded with whitish; anal tuft yellow in both sexes. Fore wings black; some bluish-white basal dots; markings orange or yellowish-white (*niveosparsa*), viz.: an antemedial cell-spot; a streak on dorsum; a triangular bar across the extremity of cell; a square spot below base of vein 2; a postmedial series of spots between veins 6 and 8, sometimes with a slight streak in interspace V, larger spots between veins 3 and 5; a spot above tornus; some silvery-blue markings, viz., above vein 1 at base, in middle of cell, on discocellulars; cilia white apically. Hind wings orange red or scarlet, with a black terminal band, running on costa to before middle; sometimes a red spot on termen in interspace I; cilia white apically. Length of fore wing 32–35 mm.

D i s t r i b u t i o n : West Africa: Gold Coast, Nigeria, Cameroon, Ashanti, Old Calabar, Ivory Coast.

Heraclia xanthopyga (MABILLE)

1890 *Eusemia xanthopyga* MABILLE, Bull. Soc. entomol. France, Paris, **90**:124. – T e r r a t y p i c a : Afrique orientale (Type, not designated; lectotype designated by VIETTE & FLETCHER 1968, ♂, in British Museum).
1892 *Xanthospilopteryx xanthopyga* – KIRBY, Syn. Cat. Lepidopt. Het.:899.
1901 *Rothia xanthopyga* – HAMPSON, Cat. Lepidopt. Phal. Brit. Mus., **3**:577.

D i a g n o s i s : Head and thorax black, the two first palpal joints, sides of frons, vertex, collar, tegulae and thorax above spotted with yellowish; fore coxae, femora, tibiae and tarsi banded orange; abdomen black, ringed with orange yellow above, and sometimes verntrally; anal tuft orange in both sexes. Fore wings black; markings orange yellow, viz., a few basal points; antemedial spots in and below cell; a squarish spot in cell-extremity, conjoined to a spot below base of vein 1; a spot on middle of dorsum; an oblique postmedial band between veins 3 and 8; a rather large spot above tornus, some silvery blue marks, biz., near base, in middle of cell, a discocellular bar, a streak below middle of cell to dorsum, cilia yellowish apically. Hind wings carmine-red with a black terminal band, running on costa to near middle and narrowing to a point at tornus, cilia yellowish apically. Length of fore wing 25–27 mm.

D i s t r i b u t i o n : Tanzania, Kenya, Uganda.

Heraclia terminatis (WALKER)

1856 *Eusemia terminatis* WALKER, List Lepidopt. Het. Brit. Mus., 7:1587. – T e r r a t y p i c a : Sierra Leone (Types, ♂, ♀, in British Museum).
1891 *Xanthospilopteryx terminatis* – KIRBY, Trans. r. entomol. Soc. London, **1891**:284.
1892 *Anaphela terminatis* – KIRBY, Syn. Cat. Lepidopt. Het.:25.

D i a g n o s i s : Head and thorax black; sides of palpi and of frons, occiput, collar spotted with white; coxae, streaks on femora and rings on tarsi white; tibiae spotted with orange; abdomen black, with white segmental bands tinged yellowish on basal half above; underside almost wholly yellow in ♂ and barred with white in ♀; extremity yellow. Fore wings black; several bluish-white basal dots; markings white, viz., an antemedial triangular cell-spot; a bar across the cell-extremity; a square spot below base of vein 2; elongate spots postmedially above veins 4, 6 and 7, with a triangular spot in interspace III; a spot above tornus, some silvery-blue marks, viz., at base above vein 1, in cell-middle, on discocellulars, and a streak in middle of the submedian fold; cilia white apically. Hind wings orange yellow with a black terminal band, running on costa to base; cilia white apically. Length of fore wing 20--24 mm.

Distribution: West Africa (Sierra Leone, Old Calabar, Cameroon to Niger).

Heraclia atrifusa (HAMPSON)

1912 *Xanthospilopteryx atrifusa* HAMPSON, Ann. Mag. nat. Hist., London, (8) **10**:610. – Terra typica: Semliki Plains, Uganda (Type, ♂, in British Museum).
1920 *Heraclia atrifusa* HAMPSON, Cat. Lepidopt. Phal. Brit. Mus., **2**:550; pl. 70, fig. 19.

Diagnosis: Head and thorax black; the two first palpal joints, sides of frons, vertex, collar, tergulae with white points; pectus and femora marked with white; tibiae spotted with orange; tarsi brown; abdomen black with narrow white rings, the second and third urites with some yellow at sides; anal tuft black; markings white, viz., a point at base below cell; a small triangular spot in base of cell; a slight spot above vein 1; a short streak above middle of dorsum; a square patch in cell-end; and oblique square patch below cell; a postmedial trifid patch below costa, divided by black streaks on veins 6 and 7; an elongate spot above vein 4; a spot in interspace III, a small spot above tornus; some silvery-blue markings, viz., some scales in basal area, an oblique streak in cell-middle, a faint streak below middle of cell, and a line on the discocellulars; cilia white apically. Hind wings pale yellow, suffused with black; a black terminal band, broad costally, and narrowing to a point towards the anal angle; cilia white apically. Length of fore wing 31 mm.

Distribution: Uganda.

Heraclia deficiens (MABILLE & VUILLOT)

1892 *Eusemia deficiens* MABILLE & VUILLOT, Novit. Lepidopt., **7**:53; pl. 8, fig. 4 – Terra typica: Afrique occidentale (Type, ♀, in British Museum).
1892 *Xanthospilopteryx deficiens* – KIRBY, Syn. Cat. Lepidopt. Het.:899.
1892 *Xanthospilopteryx minerva* WEYMER, Stettiner entomol. Ztg., **53**:103. – Terra typica: Gaboon (Type, ♂, in Berlin Museum).
1892 *Eusemia pallida* – MABILLE & VUILLOT (nec WALKER 1854), Novit. Lepidopt., **7**:53; pl. 8, fig. 4.
1913 *Xanthospilopteryx deficiens* forma *minerva* – JORDAN in SEITZ, Großschmetterl. Erde, **15**:6; pl. 1e.

Diagnosis: Head and thorax black, sides of palpi and of frons, occiput, collar, tegulae spotted with white, streaks on coxae and femora, and rings on tarsi white, tibiae spotted with orange; abdomen black, ringed with orange above and with white below, extremity black in both sexes. Fore wings black; about six bluish-white dots at base, markings white (forma *minerva*) or yellowish-white, viz., an antemedial cell-spot, two small spots above vein 1; a bar across the cell-extremity; a square spot below base of vein 2; an oblique postmedial band between veins 3 and 8; spot in interspace III more or less isolated, a spot above tornus; some silvery-blue markings, viz., at base above vein 1, in middle of cell; on discocellulars; cilia white apically.

Hind wings red with a black terminal band, occupying also the costa to near middle; cilia white apically. Length of fore wing 35 mm.

D i s t r i b u t i o n : There is some confusion about the distribution of *H. deficiens*: the typical specimen said by MABILLE to inhabit West Africa does not differ from specimens from Tanzania and from the island Pemba (in ROTHSCHILD collection, with some doubt); forma *minerva* was described from Gaboon, but STRAND in Lepidopt. Cat. mentions East Africa. VIETTE & FLETCHER (Bull. Brit. Mus. (Nat. Hist.), Entomology, **21**, (8):416) mention "Afrique occidentale" for the unique type, a ♀ without abdomen.

Heraclia medeba (DRUCE)

1880 *Eusemia medeba* DRUCE, Entomologist's mon. Mag., **16**:269. – T e r r a t y p i c a : Isubu, Old Calabar (Type, ♀, in British Museum).
1892 *Anaphela medeba* – KIRBY, Syn. Cat. Lepidopt. Het.:25.
1913 *Xanthospilopteryx medeba* – JORDAN in SEITZ, Großschmetterl. Erde, **15**:6; pl. 000.

D i a g n o s i s : Head and thorax black; head with four white spots; collar with four white spots; tegulae with a large median white spot; thorax above slightly variegated with white at middle; pectus variegated black and white; pilosity of legs yellowish, mid and hind tarsi with white spots on joints; abdomen black, with the urites finely edged with white, broader laterally, and very broad ventrally; anal tuft orange. Fore wings black markings white, viz., a triangular spot in basal part of cell; a squarish spot in distal part of cell; an elongate spot on middle of dorsum; a large double spot in interspaces I–II below the discocellulars; a postmedial series of spots in interspaces III to VII, spot in interspace V minute, that in interspace IV the largest, a small spot above tornus; some metallic blue marks, viz., two subbasal dots, two just above the anal vein, a cell-spot between the two white cell-spot, and a bar on the discocellulars; cilia white apically. Hind wings creamy white, strongly sprinkled with blackish brown, except on anal area; base, broad costal region and terminal band about 5 mm wide, dark brownish black; veins 2, 3, 4 in the pale area blackish brown; cilia white apically. Length of fore wing 31 mm.

D i s t r i b u t i o n : Old Calabar; Gaboon; Western Zaire (upper Kasai: Luluabourg = Kananga).

Heraclia contigua (WALKER)

1854 *Eusemia contigua* WALKER, List Lepidopt. Het. Brit. Mus., **1**:50. – T e r r a t y p i c a : ignota (Type, ♂, in British Museum).
1891 *Xanthospilopteryx contigua* – KIRBY, Trans. r. entomol. Soc. London, **1891**:286–287.
1901 *Rothia contigua* – HAMPSON, Cat. Lepidopt. Phal. Brit. Mus., **3**:578.

D i a g n o s i s : Body black; the two first palpal joints with white hair; sides of frons, collar, tegulae, thorax above and legs spotted with white; hind tibiae fringed with yellowish hair; abdomen with segmental yellowish-grey rings. Fore wings blackish brown; three white basal points; markings yellow, viz., a very oblique antemedial band from the subcostalis to above vein 1; a small spot in upper cell-extremity; an oblique postmedial band between veins 2 and 8; a small spot above tornus;

some silvery-blue marks, viz., two basal points, an oblique line in middle of cell, a discocellular line, a streak on vein 2 to beyond middle, and a postmedial streak on vein 1; cilia ochreous apically. Hind wings carmine-red, with a rather broad blackish-brown terminal band; cilia spotted with black. Length of fore wing 24 mm.

Distribution: Unknown. The single specimen in British Museum does not bear any indication.

Heraclia aisha (KIRBY)

1891 *Xanthospilopteryx aisha* KIRBY, Trans. r. entomol. Soc. London, **1891**:291; pl. 15, fig. 3. – Terra typica: Masailand (Type, ♀, in British Museum).
1898 *Xanthospilopteryx catarhodia* HOLLAND, in Donaldson Smith's Unknown African Countries: 409; fig. 7. – Terra typica: Dabulli (Type, ♂, in Carnegie Museum).
1901 *Rothia aisha* – HAMPSON, Cat. Lepidopt. Phal. Brit. Mus., **3**:576.
1913 *Xanthospilopteryc aisha* forma *transiens* JORDAN in SEITZ, Großschmetterl. Erde, **15**:7. – Terra typica: not designated (Type in British Museum).

Diagnosis: Body and fore wings black; sides of palpi and of frons, occiput, collar, tegulae, meso- and metathorax spotted with ochreous; coxae, tibiae and tarsi banded orange; abdomen with dorsal and lateral bands, sometimes also on the underside; anal tuft orange. Fore wings marked with orange, viz., three points below base of costa; two points below cell; one above vein 1; an antemedial cell-spot; an antemedial cell-spot; an antemedial spot above vein 1; a large spot in cell-extremity; a square spot below base of vein 2; a small spot on middle of dorsum; an oblique band beyond cell between veins 3 and 8, constricted in middle; some silvery-blue markings, viz., in and below cell at base, a bar in middle of cell, another on the discocellulars, a streak below middle of cell and postmedial spots on veins 1 and 2; cilia yellowish apically. Hind wings scarlet; some black scales at the base of the anal margin; a terminal black band, running inwards on costa to near middle; cilia orange apically. Length of fore wing 27 mm.

Distribution: East Africa: Masailand, Kenya, Uganda.

Heraclia thruppi (BUTLER)

1886 *Eusemia thruppi* BUTLER, Proc. zool. Soc. London, **1885**:775. – Terra typica: Somaliland (Type, ♀, in British Museum).
1891 *Xanthospilopteryx thruppi* – KIRBY, Trans. r. entomol. Soc. London, **1891**:289–290; pl. 15, fig. 5.
1901 *Rothia trhruppi* HAMPSON (ex errore pro *thruppi*), Cat. Lepidopt. Phal. Brit. Mus., 0000:575; pl. 53, fig. 9.

Diagnosis: Head and thorax black; palpi and sides of frons spotted with white; occiput, collar, tegulae, meso- and metathorax spotted with yellow; coxae, and tibiae spotted with orange; spurs white; tarsi ringed with white; abdomen reddish above, black below; two dorsal black spots on the basal urite, and a lateral series of black spots. Fore wing black; markings yellow, viz., several basal points; an antemedial spot in cell; a triangular spot on the dorsal area; a square patch in extremity of cell, with a streak from its inner side below costa; a rounded spot below base of vein 2; an oblique postmedial patch between veins 3 and 8, rounded at the extremities; a large spot above tornus; some silvery-blue marks, viz. scales on the basal area, below middle and extremity of cell, bars across middle of cell and on the discocellulars, and spots on veins 1 and 2. Hind wings carmine-red, becoming orange towards the termen; terminal black band rather narrow, running on costa towards near middle. Length of fore wing 40 mm.

Distribution: East Africa: Kenya, Uganda, Somaliland.

Heraclia durbanica (STONEHAM)

1963 *Xanthospilopteryx durbanica* STONEHAM, Bull. Stoneham Museum Nairobi, **79**:1. – Terra typica: Durban, Natal (Type in Stoneham Museum Nairobi).

Diagnosis: "Very near *X. thruppi*, but the abdomen is not black beneath. There are white spots on the palpae (sic) and at the basis of the antennae. On the thorax are four white spots in a line, and two large shoulder-spots. On the abdomen at the junction with the thorax is a dorsal yellow tuft".

Distribution: Natal. This species is unknown to the writer.

Heraclia hornimani (DRUCE)

1880 *Eusemia hornimani* DRUCE, Entomologists's mon. Mag., **16**:269. – Terra typica: Mongo-ma-Lobah, West Africa (Type, ♀, in British Museum).
1892 *Xanthospilopteryx hornimanni* – KIRBY (ex errore pro *hornimani*), Syn. Cat. Lepidopt. Het.:30.
1892 *Xanthospilopteryx hornimani* – KIRBY, Syn. Cat. Lepidopt. Het.:898.
1920 *Heraclia hornimani* – HAMPSON, Cat. Lepidopt. Phal. Brit. Mus., Suppl. **2**:561.

Diagnosis: Body black; the first two palpal joints, sides of frons, occiput, collar, tegulae, pectus and legs with white points; tibiae with some orange; abdomen with some orange hair at base, and with dorsal, lateral, sublateral and ventral series of white points, anal tuft orange in ♀. Fore wings black; markings orange in ♂, white or yellowish white in ♀, viz., a few points at the base; a wedge-shaped spot in cell, an elliptical patch below cell, a square spot in cell-extremity; an elongate or wedge-shaped spot below vein 2, a series of elongate spots beyond the cell, between veins 2 or 3 and 8, sometimes fused to a band; a spot above tornus; some silvery-blue in and below middle of cell, and on the discocellulars. Hind wings orange red or yellowish white, with a terminal black band extending on costa to near base, and narrowing to a point before the anal angle. Length of fore wing 27–41 mm.

Distribution: West, Central and East Africa.

Heraclia hornimani hornimani (DRUCE)

1910 *Xanthospilopteryx comaria* DRUCE, Ann. Mag. nat. Hist., London, (8) **5**:395. – Terra typica: Ja River, Bitje, Cameroon (Type, ♂, in British Museum).
1913 *Xanthospilopteryx hornimani* ssp. *hornimani* – JORDAN in SEITZ, Großschmetterl. Erde, **15**:pl. 2c and pl. 4f.

D i a g n o s i s : ♀ with the markings pure white.

D i s t r i b u t i o n : West Africa: Cameroon, Sierra Leone, Nigeria, Gold Coast, Gaboon; South Zaire: Lulua, Shaba.

Heraclia hornimani melanosoma (DRUCE)

1901 *Xanthospilopteryx melanosoma* DRUCE, Ann. Mag. nat. Hist., London, (7) **7**:433. – T e r r a t y p i c a : East Africa (Type, ♂, in British Museum).
1907 *Xanthospilopteryx minchini* HAMPSON, Ann. Mag. nat. Hist., London, (7) **19**:241. – T e r r a t y p i c a : Kavirondo (Type, ♀, in British Museum).
1913 *Xanthospilopteryx hornimani* ssp. *melanosoma* – JORDAN in SEITZ, Großschmetterl. Erde, **15**:5; pl. 1 e.
1913 *Xanthospilopteryx hornimani* aberr. *minchini* – JORDAN in SEITZ, Großschmetterl. Erde, **15**:5.

D i a g n o s i s : Fore wings of ♂ with the markings deeper orange, those of ♀ yellowish white, broader, especially in the form *minchini*, where they are confluent; sometimes most markings are diffuse, and the hind wings whitish proximally, and brown distally (form *diffusa*).

D i s t r i b u t i o n : East Africa: Uganda; North and South Zaire: Ituri, Kivu, Lulua; Burundi; Rwanda.

Heraclia hornimani meridionalis KIRIAKOFF

1955 *Heraclia hornimani meridionalis* KIRIAKOFF, Bull. Inst. r. Sci. nat. Belg., Bruxelles, **31** (69):6–7; fig. 3. – T e r r a t y p i c a : Haut Katanga (Type, ♂, in Brussels Museum).

D i a g n o s i s : On the fore wings, the basal triangular spot, the two medial spots and the oblique postmedial band are much reduced; hind wings orange costally, red elsewhere, with the terminal band rather narrow.

D i s t r i b u t i o n : Extreme South of Zaire: Shaba.

Heraclia hypercompoides (BUTLER)

1895 *Xanthospilopteryx hypercompoides* BUTLER, Proc. zool. Soc. London, **1895**:739; pl. 43, fig. 5. – T e r r a t y p i c a : Ruwenzori, Kenya (Type, ♀, in British Museum).

D i a g n o s i s : Body black; sides of palpi and of frons, occiput, collar, tegulae spotted with ochreous-white; coxae, tibiae and tarsi spotted with white; tibiae fringed with orange hair on outer side, abdomen with lateral and ventral series of white dots, extremity orange in ♀. Fore wings black; markings yellow, viz., some basal dots below costa to the submedial fold, with a rounded spot below it, a bar from below vein 2; a trifid band beyond cell from below costa to vein 5, with a wedge-shaped patch between veins 3 and 5, extending nearly to termen; a spot above tornus. Hind wings orange-red or red, with a terminal black band narrowing towards the anal angle, but very broad apically, and extending on costa to before middle. Length of fore wing 25–30 mm.

D i s t r i b u t i o n : Kenya, Uganda, Kivu region.

Heracila elongata (BARTEL)

1903 *Xanthospilopteryx elongata* BARTEL, Verh. zool.-bot. Ges. Wien, **53**:124. – Terra typica: Manow, Deutsch Ostafrika = Malawi (Type, ♂, in British Museum).
1920 *Rothia elongata* – HAMPSON, Cat. Lepidopt. Phal. Brit. Mus., Suppl. **2**:565.

Diagnosis: Body blackish-brown; the two first palpal joints spotted with white; sides of frons, occiput, collar, shoulders spotted with orange yellow; bases of tegulae with some orange yellow scales; cheeks yellow; coxae spotted with white; tibiae banded yellow, with the extremities white; tarsi ringed with white, abdomen with small white lateral spots except at base, and with yellow sublateral streaks towards base. Fore wings black; markings orange yellow, viz., two spots below base of costa, a small subbasal spot above vein 1; an oblique antemedial band from below costa to vein 1; a square spot towards the cell-extremity; an elliptical spot below vein 2; an oblique band from below costa beyond middle to below vein 4 near termen where its extremity is hooked; a small subterminal spot in the submedial fold; silvery-blue markings, viz., some sprinkling on basal area, in cell-middle, on discocellulars. Hind wings orange-red, with a rather narrow terminal black band, angled at vein 4. Length of fore wing 25–29 mm.

Distribution: Malawi. A single specimen is known, ex ROTHSCHILD collection, in British Museum.

Heraclia zenkeri (KARSCH)

1895 *Xanthospilopteryx zenkeri* KARSCH, Entomol. Nachr.Bl., Wien, **21**:350; pl. 2, fig. 1. – Terra typica: Cameroon (Type, ♀, in Berlin Museum).
1901 *Xanthospilopteryx zenceri* (HAMPSON (ex errore pro *zenkeri*), Cat. Lepidopt. Phal. Brit. Mus., **3**:568–569.
1920 *Heraclia zenceri* – HAMPSON, Cat. Lepidopt. Phal. Brit. Mus., Suppl. **2**:552.

Diagnosis: Head and thorax black; vertex and collar spotted with bluish, tegulae and metathorax with short bluish streaks; abdomen orange, sometimes more or less tinged blackish, a dorsal basal blackish mark; extremity blackish. Fore wings orange with a greenish tinge, and the base blackish; a blackish costal fascia, with four bluish-white spots basally; an oblique elliptical black medial spot in and above cell, with blue centre; terminal area black from costa just beyond middle, with irregular proximal margin; a blue lunule on the discocellulars; area between the elliptical medial spot and the terminal area orange, from below costa to below vein 3; a small orange spot above tornus; cilia white apically. Hind wings orange-red, with a black terminal band, broad costally, and narrowing to a point at tornus; cilia white. Length of fore wing 28–42 mm.

Distribution: Most of tropical Africa, from Cameroon to Sudan and Tanzania.

Heraclia zenkeri zenkeri (KARSCH)

1913 *Xanthospilopteryx zenkeri* ssp. *zenkeri* – JORDAN in SEITZ, Großschmetterl. Erde, **15**:2; pl. 000.
1913 *Xanthospilopteryx zenkeri* forma *flaviventris* – JORDAN in SEITZ, Großschmetterl. Erde, **15**:2, pl. 2b. (as *zenkeri*).

1913 *Xanthospilopteryx zenkeri* forma *nigriventris* JORDAN in SEITZ, Großschmetterl. Erde, **15**:2; pl. 000. – Terra typica: not designated (Type in British Museum).

Diagnosis: Ground colour of fore wings varying from ochreous yellow to reddish brown; hind wings yellow or orange. Abdomen of ♀ varying from pure yellow above, and black faintly banded with yellow below (forma *flaviventris*).

Distribution: Cameroon.

Heraclia zenkeri zeodita (NIEPELT)

1901 *Xanthospilopteryx ochracea* AURIVILLIUS, Entomol. Tidskr., Stockholm, **22**:122 (nec *Eusemia ochracea* BUTLER 1975 = *Eusemia pardalina* WALKER 1869). – Terra typica: Congo (Type, ♀, in Museum Stockholm).

1906 *Xanthospilopteryx zeodita* NIEPELT, Insekten-Börse, **23**:127. – Terra typica: Kasai (Type in coll. NIEPELT; Syntypes in British Museum).

1913 *Xanthospilopteryx zenkeri* ssp. *zeodita* – JORDAN in SEITZ, Großschmetterl. Erde, **15**:2.

Diagnosis: Fore wings with the cilia white apically and tornally; in ♂ with two antemedial yellow spots, in ♀ with the discal black band broader and connected with the black costal fascia; abdomen varying from yellow to nearly pure black.

Distribution: Most of Zaire.

Heraclia zenkeri niepelti (OBERTHÜR)

1911 *Xanthospilopteryx niepelti* OBERTHÜR, Bull. Soc. entomol. France, Paris, **1910**:472; pl. 16, fig. 2. – Terra typica: M'pala, Lake Tanganyika region, Congo (Type, ♀, in British Museum).

1913 *Xanthospilopteryx zenkeri* ssp. *niepelti* – JORDAN in SEITZ, Großschmetterl. Erde, **15**:3.

1920 *Heraclia niepelti* – HAMPSON, Cat. Lepidopt. Phal. Brit. Mus., Suppl. **2**: 553.

Diagnosis: Fore wings with the cilia white apically and tornally; the lower antemedial spot broadly connected with the terminal band, and through a black line below cell with the discal band, so that there remains an isolated medial orange patch; terminal black band broader, and with orange not white spots.

Distribution: Eastern Zaire.

Heraclia karschi (HOLLAND)

1892 ? *Xanthospilopteryx abacata* KARSCH, Entomol. Nachr.Bl., Wien, **18**:181. Terra typica: Deutsch-Südwestafrika [errore: Cameroon, fide JORDAN 1913] (Type formerly in Museum Berlin).

1892 ? *Xanthospilopteryx lomata* KARSCH, Entomol. Nachr.Bl., Wien, **18**:182. Terra typica: Deutsch-Südwestafrika [errore: Cameroon, fide JORDAN 1913] (Type formerly in Museum Berlin).

1897 *Xanthospilopteryx karschii* HOLLAND, Ann. mag. nat. Hist., London (6) **20**:292. – Terra typica: Cameroons (Type, ♀, in Museum Berlin).

1897 *Xanthospilopteryx kirbyi* HOLLAND, Ann. Mag. nat. Hist., London, (6) **20**:292 (nec *Xanthospilopteryx kirbyi* CARPENTER 1893). – Terra typica: Equatorial Africa. (Type, ♂, in Carnegie Museum).

1901 *Xanthospilopteryx nigridorsa* aberr. *karschi* – HAMPSON, Cat. Lepidopt. Phal Brit. Mus., **3**:572.
1901 *Xanthospilopteryx cirbyi* HAMPSON (ex errore pro *kirbyi*), Cat. Lepidopt. Phal. Brit. Mus., **3**:569; pl. 52, fig. 17.
1903 *Xanthospilopteryx triseriata* BARTEL, Verh. zool.-bot. Ges. Wien, **53**:125. – Terra typica: Yaonde, Cameroon (Type, ♀, in Museum Berlin).
1913 *Xanthospilopteryx hollandi* JORDAN (f. of *karschi* HOLLAND) in SEITZ, Großschmetterl. Erde, **15**:3
1908 *Xanthospilopteryx egregia* WICHGRAF, Internation. entomol. Z., Guben, **22**:106. – Terra typica: Cameroon (Type, ♀, in British Museum).
1912 *Xanthospilopteryx tessmanni* STRAND, Arch. Naturgesch., Berlin, **78** A (6):195. – Terra typica: Cameroon (Type, ♀, in Museum Berlin).
1913 *Xanthospilopteryx karschi* forma *karschi* – JORDAN in SEITZ, Großschmetterl. Erde, **13**:3; pl. 2c.
1920 *Heraclia cirbyi* – HAMPSON, Cat. Lepidopt. Phal. Brit. Mus., Suppl. **2**:557.

Diagnosis: Head and thorax black; first and second palpal joints, sides of frons, occiput, with white points; collar with four pale yellow spots; streaks on femora and rings on tarsi white; tibiae fringed with orange hair on outer side; abdomen yellow or blackish, or mixed yellow and black, sometimes spotted with white or yellow below; extremity often orange in ♀. Fore wings black; three small yellow basal spots, ochreous yellow markings, viz., a medial band of two large spots separated by the black median vein, spot below the latter more or less lunulate; an oblique mark on dorsal area; an oblique postmedial band of four elongate spots, from below costa to vein 4, an elongate spot in interspace III at an acute angle with the spot in interspace IV, a spot above tornus; a few silvery-blue dots in cell before the band, and on the discocellulars; cilia white apically. Hind wings red, orange or yellow; basal and anal areas with a blackish suffusion; a broad terminal black band, broader in the apical area, and reaching the middle of costa; cilia white at the extremity. Length of fore wing 28--45 mm.

Distribution: Cameroon to Uganda and Zaire (not yet found in Shaba).

Heraclia gruenbergi (WICHGRAF)

1911 *Xanthospilopteryx gruenbergi* WICHGRAF, Internation. entomol. Z., Guben, **5**:175. – Terra typica: Nairobi, Kenya (Type, ♂, in British Museum).
1912 *Xanthospilopteryx atribasalis* HAMPSON, Ann. Mag. nat. Hist., London, (000) **10**:610. – Terra typica: Yala River, Kakumega Forest, East Africa (Type, ♂, in British Museum).
1920 *Heraclia atribasalis* – HAMPSON, Cat. Lepidopt. Phal. Brit. Mus., Suppl. **2**:556; pl. 70, fig. 20.
1920 *Xanthospilopteryx gruenbergi* – HAMPSON, Cat. Lepidopt. Phal. Brit. Mus., Suppl. **2**:560.

Diagnosis: Head and thorax black; first and second palpal joints, sides of frons, occiput, collar and tegulae with white dots; legs black variegated with white; pilosity of fore tibiae partly orange; abdomen yellowish mixed black, or black, sometimes yellowish below, and with the extremity orange in ♀. Fore wings black; sometimes a few small witish basal dots; markings creamy white, viz., an oblique medial band consisting of a large square cell-patch, and a more or less triangular patch below cell; a postmedial band of spots, from below costa to vein 5, below vein 5, elongate, and a spot below the extremity of the preceding; a subterminal dot in the submedian fold, silvery-blue spots at the base a streak in middle of cell, and on the discocellulars; cilia white apically. Hind wings creamy-white (*X. gruenbergi*) or orange (*X. atribasalis*), base and anal margin brownish black; a terminal black band, broad costally and narrowing to tornus; cilia white apically. Length of fore wing 29–44 mm.

Distribution: Uganda to most of Zaire.

Heraclia gruenbergi gruenbergi (WICHGRAF)

Diagnosis: Fore wings without dorsal spot.

Distribution: Uganda.

Heraclia gruenbergi medjensis (HOLLAND)

1920 *Xanthospilopteryx medjensis* HOLLAND, Bull. amer. Mus. nat. Hist., New York, **43**:271; pl. 13, fig. 1. – Terra typica: Medje, Uele (Type, ♂, in American Museum of natural History).
1934 *Xanthospilopteryx gruenbergi medjensis* – GAEDE in SEITZ, Großschmetterl. Erde, **15**:25.

Diagnosis: Fore wings with a large triangular dorsal spot; hind wings yellowish white.

Distribution: Zaire (Uele, Kisangani District, Lulua).

Heraclia flavipennis (BARTEL)

1903 *Xanthospilopteryx flavipennis* BARTEL, Verh. zool-bot. Ges. Wien, **53**:121. – Terra typica: "West Africa" [Congo region, fide JORDAN 1913] (Type, ♀, in British Museum).
1920 *Heraclia flavipennis* – HAMPSON, Cat. Lepidopt. Phal. Brit. Mus., Suppl. **2**:558.

Diagnosis: Head and thorax black; first and second palpal joints, sides of frons, occiput, collar and tegulae with white spots, sides of head white; pectus, coxae, tibiae and tarsi marked with white; abdomen yellow, slightly tinged with black; two basal bands and extremity black. Fore wings black; markings yellow; a triangular antemedial cell-patch; almost confluent fasciae below cell below and and above the submedial fold, and above dorsum where the yellow colour mostly extends from the base to well beyond the middle; an oblique conical patch in cell-end, an oblique patch below the basal half of vein 2; an oblique postmedial patch from below costa to below vein 3 near termen, its inner edge slightly irregular; a small subterminal spot below vein 2, silvery-blue bar in middle of cell, and a discocellular lunule; cilia white apically. Hind wings orange or reddish orange; a black terminal band extending towards the apex on vein 2; cilia white apically, and with white extremities from vein 3 to the anal angle. Length of fore wing 29–41 mm.

Distribution: Lualaba, Shaba, East Zaire (Semliki valley).

Heracliа longipennis (WALKER)

1854 *Eusemia longipennis* WALKER, List Lepidopt. Het. Brit. Mus., 1:51. – Terra typica: Ashanti (Types, ♂, ♀, in British Museum).
1891 *Xanthospilopteryx longipennis* – KIRBY, Trans. r. entomol. Soc. London, **1891**:286.
1910 *Xanthospilopteryx dönitzi* GRUENBERG, S. Ber. Ges. naturf. Fr. Berlin, **1910**:176; fig. 9. – Terra typica: Sesse Island, Lake Victoria, Uganda (Type, ♀, in Berlin Museum).
1912 *Xanthospilopteryx kirbyi* aberr. *alice* STRAND, Lepidopt. Cat., **5**:16. – Terra typica: Pt. Alice (Unyoro) (aberr. 2 of Hampson, Cat. Lepidopt. Phal. Brit. Mus., **3**:570).
1920 *Heraclia longipennis* – HAMPSON, Cat. Lepidopt. Phal. Brit. Mus., Suppl. **2**:556.

Diagnosis: Head and thorax black, sides of palpi and of frons, occiput, collar and tegulae with white points; chin, coxae, streaks of femora, rings on tarsi, white, tibiae with a few orange hairs on the outer side; abdomen orange, blackish at the base, with series of lateral black spots in ♀. Fore wings black, markings orange, viz., an antemedial spot sometimes present in cell; an elongate spot on dorsum; a bar across end of cell, with a streak above it and a squarish spot in interspace I, a series of elongate spots beyond cell between veins 5 and 8; a streak above vein 4 with a more rounded spot below its extremity, a spot above tornus, silvery blue dots at the base, and some silvery blue in middle of cell and on the discocellulars; cilia white apically. Hind wings orange red with the base black; a broad terminal black band, reaching the base on costa, cilia white apically, in mostly white. Length of fore wing 38–48 mm.

Distribution: Ashanti, Gold Coast, Ivory Coast, Cameroon to northern Zaire and Uganda.

Heraclia longipennis longipennis (WALKER)

Diagnosis: Medial spots not conjoined.
Distribution: Most of the area of distribution.

Heraclia longipennis bodaensis (GAEDE)

1900 *Xanthospilopteryx longipennis bodaensis* GAEDE, Internation. entomol. Z., Guben, **9**:80. – Terra typica: Cameroon (Type, Museum Berlin).

Diagnosis: Yellow spots larger; medial spots conjoined; subapical series also continous, three small subbasal spots.

Distribution: Cameroon.

Heraclia aemulatrix (WESTWOOD)

1881 *Eusemia aemulatrix* WESTWOOD, Oates' Matabeleland:355. – Terra typica: Gold Coast (Type, ♂, in Museum Oxford University).
1892 *Xanthospilopteryx aemulatrix* – KIRBY, Syn. Cat. Lepidopt. Het.:30 and 898.
1920 *Heraclia aemulatrix* – HAMPSON, Cat. Lepidopt. Phal. Brit. Mus., Suppl. **2**:557.

D i a g n o s i s : Head and thorax black; sides of palpi and of frons, occiput, collar and tegulae with white points; chin, coxae, streaks on femora and rings on tarsi white; tibiae tinged with orange; abdomen orange, with the extremity blackish in ♂. Fore wing black, markings ochreous yellow, viz., a small triangular antemedial cell-spot; a fine streak above vein 1; a dorsal streak; a bar across the cell-extremity; a square spot below the base of vein 2; a series of oblong spots beyond cell, between veins 5 and 8, conjoined in ♀; a longer streak above vein 4, a rounded spot below its extremity, three bluish dots below costa, and two below cell; blue scales in and below middle of cell, and on the discocellulars; cilia white apically. Hind wings red, orange red or yellowish white, with a terminal black band, broader at the apex but not reaching the middle of costa, cilia white apically. Length of fore wing 23–34 mm.

D i s t r i b u t i o n : Gold Coast and Cameroon to whole of Zaire, Burundi, Uganda, Zambia.

Heraclia aemulatrix aemulatrix (WESTWOOD)

1892 *Eusemia longipennis* – MABILLE, Bull. Soc. entomol. France, Paris, **1890**:123 (nec WALKER 1854).
1903 *Xanthospilopteryx mabillei* BARTEL, Berh. zool.-bot. Ges. Wien, **53**:122. – T e r r a t y p i c a : Cameroun, Yaonde (Type, ♂, in Museum Berlin).
1913 *Xanthospilopteryx aemulatrix* ssp. *aemulatrix* – JORDAN in SEITZ, Großschmetterl. Erde, **15**:4; pl. 1f (as *mabillei*).

D i a g n o s i s : Hind wings red with the black terminal band narrow, at most 3,5 mm wide.

D i s t r i b u t i o n : Zaire, Rhodesia, Gaboon, Cameroon.

Heraclia aemulatrix campala (JORDAN)

1912 *Xanthospilopteryx kirbyi* aberr. *campala* STRAND, Lepidopt. Cat., **5**:16. – T e r r a t y p i c a : Kampala, "British Ostafrika" (Type, ♂, in British Museum).
1913 *Xanthospilopteryx aemulatrix* ssp. *campala* – JORDAN in SEITZ, Großschmetterl. Erde, **15**:4.
1913 *Xanthospilopteryx aemulatrix* ssp. *campala* forma *perpallida* JORDAN in SEITZ, Großschmetterl. Erde **15**:4.

D i a g n o s i s : Hind wings red or orange red (forma *campala*) or yellowish (forma *perpallida*) with the black terminal band broad, at least 5 mm wide in the middle portion.

D i s t r i b u t i o n : Uganda, Toro.

Heraclia aemulatrix suda (JORDAN)

1913 *Xanthospilopteryx aemulatrix* ssp. *suda* JORDAN in SEITZ, Großschmetterl. Erde, **15**:4. – T e r r a t y p i c a : Kwidgwi Island, Lake Kivu (Types, ♂, ♀, in British Museum).

D i a g n o s i s : Spots on head and thorax yellowish; wings somewhat more rounded; hind wings with the black margin mostly narrower.

D i s t r i b u t i o n : Zaire (Kivu District).

Heracliа monslunensis (HAMPSON)

1901 *Xanthospilopteryx mons-lunensis* HAMPSON, Cat. Lepidopt. Phal. Brit. Mus., **3**:570–571; pl. 53, fig. 10. – Terra typica: Ruwenzori, Kenya (Type, ♂, in British Museum).
1903 *Xanthospilopteryx signata* BARTEL, Verh. zool.-bot. Ges. Wien, **53**:122. – Terra typica: S. W. Victoria, Nyanza (Type, ♀, in Museum Berlin).
1910 *Xanthospilopteryx neavi* HAMPSON, Proc. zool. Soc. London, **1910**:397; pl. 36, fig. 14. – Terra typica: Chiusali District, Northeast Rhodesia = Zambia (Type, ♀, in British Museum).
1912 *Xanthospilopteryx monslunensis* aberr. *ruwensorica* STRAND, Lepidopt. Cat., **5**:16. – (aberr. 1 of HAMPSON, Cat. Lepidopt. Phal. Brit. Mus., **3**:571).
1920 *Heraclia neavi* – HAMPSON, Cat. Lepidopt. Phal. Brit. Mus., Suppl. **2**:558.
1920 *Xanthospilopteryx signata* – HAMPSON, Cat. Lepidopt. Phal. Brit. Mus., Suppl. **2**:558, pl. 71, fig. 1.

Diagnosis: Head and thorax black; sides of palpi and of frons spotted with white; occiput, collar and tegulae spotted with yellow; chin, coxae, streaks on femora and rings on tarsi white; tibiae spotted with orange; abdomen dirty orange with blackish dorsal pilosity and a lateral series of black spots. Fore wings black; markings yellowish, viz., several basal spots; a bar-shaped spot across the end of cell; an oblique elliptical spot below base of vein 2; a series of elongate conjoined spots beyond cell between veins 4 and 8, with a rounded spot in interspace III; sometimes a confuse spot on dorsum; a spot above tornus; silvery-blue marks in and below middle of cell and on the discocellulars; cilia yellowish apically. Hind wings ochreous yellow or scarlet; base black; a terminal black band expanding widely on apical area and on vein 2, cilia yellowish apically. Length of fore wing 34–36 mm.

Distribution: Kenya, Uganda, Southeast Sudan, Burundi, Shaba, Zambia; Cameroon (?).

Heraclia nigridorsa (MABILLE)

1890 *Eusemia nigridorsa* MABILLE, Bull. Soc. entomol. France, Paris, **1890**:123. – Terra typica: "Afrique occidentale" (Type, ♀, in Berlin Museum).
1892 *Xanthospilopteryx nigridorsa* – KIRBY, Syn. Cat. Lepidopt. Het.:898.

Diagnosis: Body black; sides of palpi and of frons, occiput, collar and tegulae with white points, chin, coxae, and streaks on femora white; tibiae with some orange, abdomen with slight orange lateral streaks on the basal segments, and white points on the last four segments; anal tuft mixed orange and black. Fore wings black; markings orange viz., an oblique bar from below costa across cell-extremity, with a wedge-shaped spot below it and vein 2; an elongate spot in middle of dorsum; elongate conjoined spots beyond cell between veins 5 and 8; a longer spot below vein 5, and below that a triangular spot; three bluish-white spots at base of costa; cilia tipped with white and white apically. Hind wings black with a crimson-red patch in middle. Length of fore wing 45 mm.

Distribution: Cameroon, Gaboon.

Heraclia kivuensis KIRIAKOFF

1973 *Heraclia kivuensis* KIRIAKOFF, Lambillionea, **72–73**:60. – Terra typica: Kwidgwi, Lake Kivu, Zaire (Type, ♀, in Paris Museum).

D i a g n o s i s : Head, body and legs black; first and second palpal joints spotted with white laterally, prae- and post-ocular region, vertex, collar and joints of the legs spotted with white; first joint of the fore tarsi partly orange; last abdominal segment with a few terminal orange hairs above. Fore wings black with snowy white markings, viz., a broad antemedial band, with fine black veins on it, separating it into three spots, an elongate costal spot; a square cell-spot, widening dorsally; a more elongate spot below cell, sending dorsally a narrow line towards the base, a dorsal spot occupying the entire middle of the dorsal area; a postmedial spot from below costa to vein 4, with fine black veins on it, separating it in spots; a spot in interspace IV, produced into a point towards the termen; a small rounded spot in interspace III; a rather diffuse small spot above tornus; blue markings viz., three subbasal spots in the costal area; a cell-spot just before the white band; a subbasal spot, just below the median vein; a fine edge to the discocellulars; cilia white apically. Hind wings black with a large cinnabarred subbasal and discal patch, 13 mm wide; base black, 4 mm wide, with a black streak along the median vein to base of vein 2; anal margin broadly black; cilia white apically. Length of fore wing 35–42 mm.

D i s t r i b u t i o n : Zaire (Lake Kivu region).

Heracliabuchholzi (PLÖTZ)

1880 *Eusemia buchholzi* PLÖTZ, Stettiner entomol. Ztg. **41**:82 – T e r r a t y p i c a : Ghana, Aburi (Type, ♂, destroyed).

1880 *Eusemia grandis* DRUCE, Entomologist's mon, Mag., **16**:268. – T e r r a t y p i c a : Mongo-ma, West africa (Type, ♂, in British Museum).

1885 *Emesia grandis* – WATERHOUSE (ex errore pro *Eusemia*), Aid to Nature, **2**:pl. 65, fig. 3 and 4.

1892 *Anaphela grandis* – KIRBY, Syn. Cat. Lepidopt. Het.:25.

1892 *Eusemia melanochiton* MABILLE, Bull. Soc. entomol. France, Paris, **1891**:184. – T e r r a t y p i c a : Barombi Station, Cameroon? (Type, ♀, in Berlin Museum).

1901 *Xanthospilopteryx grandis* – HAMPSON, Cat. Lepidopt. Phal. Brit. Mus., **3**:572.

1910 *Xanthospilopteryx batesi* DRUCE, Ann. Mag. nat. Hist., London, (8) **5**:395. – T e r r a t y p i c a : Bitje, Ja River, Cameroon (Type, ♀, in British Museum).

1913 *Xanthospilopteryx grandis* aberr. *batesi* – JORDAN in SEITZ, Großschmetterl. Erde, **15**:5; pl. 4f.

1920 *Heraclia batesi* – HAMPSON, Cat. Lepidopt. Phal. Brit. Mus., Suppl. **2**:560.

1923 *Xanthospilopteryx grandis schoutedeni* HULSTAERT, Rev. Zool. Bot. Afr., Bruxelles, **11**:202. – T e r r a t y p i c a : "Congo" (Type, ♀, in Museum Tervuren).

1923 *Xanthospilopteryx grandis subfascia* HULSTAERT, Rev. Zool. Bot. Afr., Bruxelles, **11**: 201. – T e r r a t y p i c a : "Congo" (Type, ♀, in Museum Tervuren).

1924 *Heraclia barnsi* JOICEY & TALBOT, Bull Hill Museum Witley, **1**:544–555. – T e r r a t y p i c a : Lowa Valley, Northwest Kivu (Type, ♀, in British Museum).

1934 *Xanthospilopteryx buchholzi* forma *barnsi* – GAEDE in SEITZ, Großschmetterl. Erde, **15**:25.

D i a g n o s i s : Body and wings black; sides of palpi and of frons, occiput, collar and tegulae with white points; chin, coxae, streaks on femora white; tibiae marked with orange; abdomen with the anal tuft mixed orange. Fore wings with three bluish-white basal spots and ochreous yellow markings, viz., an oblique bar from below costa across the cell-extremity; a wedge-shaped spot below it and vein 2; sometimes an elongate patch on dorsum (aberr. *batesi*); elongate conjoined spots beyond cell between veins 5 and 8; a longer spot above vein 4, with a triangular spot below it; a spot above tornus, some silvery-blue in middle of cell and on the discocellulars;

cilia white apically. Hind wings with a red or yellow patch on underside only. Length of fore wing 28–47 mm.

D i s t r i b u t i o n : Cameroon to Zaire (Uele, Kinshasa, Kisangani, Kivu).

Heraclia annulata (AURIVILLIUS)

1925 *Xanthospilopteryx annulata* AURIVILLIUS, Ergebn. 2. Dtsch. Zentral-Afr. Exp., **2** (18): 1309. – T e r r a t y p i c a : Congo: Kimuenza (?) (Type, ♀, in Stockholm Museum).

D i a g n o s i s : Body black, head and thorax spotted with white; anal tuft with some yellow scales below. Fore wings black with a large medial golden yellow area, not reaching costa nor tornus, nearly straight proximally, and enclosing a large black transversal patch with a few blue scales on the discocellulars; a small yellow spot near tornus; basal area with a few blue scales and three blue points on costa, cilia blackish, but white apically. Hind wings uniform black; cilia black basally, then white, and white apically. Length of fore wing 26 mm.

D i s t r i b u t i o n : Zaire (only the type specimen known).

Heraclia aurea (WICHGRAF)

1918 *Xanthospilopteryx aurea* WICHGRAF, Internation. entomol. Z., Guben, **12** (4): 26. – T e r r a t y p i c a : Angola: Bailundu (Type, ?, in Zool. Museum München).

D i a g n o s i s : Head and thorax black; two bluish dots above the eyes; collar with two white, and between these two blue dots; thorax with four bluish tufts, and one such on each tegula; abdomen light golden brown, with black pilosity on the second and last third; anal tuft black; legs golden brown. Fore wings black; markings lemon-yellow, viz., a triangular cell-spot, reaching dorsally a spot in interspace Ib, which last is twice as long and arched above; a spot in interspace Ia parallel with dorsum, more or less straight above, and separated from the preceding spot by the dark vein; two large medial cell-spots and in interspaces Ia and I; five discal spots fused together, and connex distally, notched proximally by the black veins. Hind wings orange yellow without dark at base; terminal black band wider than 5 mm, reaching costally to two-thirds; cilia white apically and tornally in both wings. Length of fore wing 34 mm.

D i s t r i b u t i o n : Angola.

Heraclia poggei (DEWITZ)

1879 *Eusemia poggei* DEWITZ, Mitt. Münchener entomol. Ges., **3**:31; pl. 2, fig. 3. – T e r r a t y p i c a : Guinea (Type, ♂, in ? Museum Berlin).

1891 *Eusemia falkensteini* DEWITZ, Nova Acta Acad. Leop.-Carol., Halle, **42**:66; pl. 2, fig. 10. – T e r r a t y p i c a : Chinchoxo (Type, ♀, in Museum Berlin).

1892 *Xanthospilopteryx poggei* – KIRBY, Syn. Cat. Lepidopt. Het.:29.

1963 *Xanthospilopteryx poggei* forma *duplicata* STONEHAM, Bull. Stoneham Mus. Nairobi, 79:1. – T e r r a t y p i c a : Not indicated (Type, in Stoneham Museum Nairobi).

1963 *Xanthospilopteryx poggei* forma *partita* STONEHAM, Bull. Stoneham Mus. Nairobi, 79:1. – T e r r a t y p i c a : Not indicated (Type, in Stoneham Museum Nairobi).

D i a g n o s i s : Head and thorax black; first and second palpal joints, sides of frons, occiput, collar and tegulae spotted with white; chin white; coxae, femora, extremities of tibiae and the first tarsal joint spotted with white; medial part of tibiae orange above; abdomen black, but orange dorsally from base to end of the 6th urite, with black bands in ♀; anal tuft black tipped with white. Fore wings reddish-orange; extreme base and a costal line black; two white and one silvery-blue basal spots; dorsum black except basally; at middle of cell, wedge-shaped black spots marked with silvery-blue; a discocellular mark of black with a silvery-blue bar on it, both absent in ♀; a terminal black band, very broad apically and broader at tornus, with a series elongate white spots between veins 4 and 8; above tornus a white spot and two silvery-blue marks; cilia spotted with white. Hind wings orange-red; terminal band black, angled inwards on vein 2; cilia spotted white and black. Length of fore wing 24–36 mm.

R e m a r k s : Mimetic, copying the Hypsid *Helcita* and the Geometrid *Aletis*.

D i s t r i b u t i o n : Cameroon and Nigeria to Uganda, whole of Zaire including the Kivu region, and North Angola.

Heraclia catori (JORDAN)

1904 *Xanthospilopteryx catori* JORDAN, Novit. Zool., **11**:443. – T e r r a t y p i c a : Moyamba, Sierra Leone (Types, ♂, ♀, in British Museum).
1920 *Heraclia catori* – HAMPSON, Cat. Lepidopt. Phal. Brit. Mus., Suppl. **2**:553.

D i a g n o s i s : Dimorphic. In ♂ head and thorax black; first and second palpal joints, sides of frons, basal antennal joints dotted with white; collar, tegulae and thorax above spotted with pale yellow; pectus and legs spotted with white; tibiae banded orange; abdomen black, but orange dorsally on the last urites, and with sublateral, lateral and ventral series of white spots. Fore wings black; markings orange, viz., a wedge-shaped patch from base to before middle, produced on vein 1 to a point beyond the middle; an oblique medial band from the subcostal vein to vein 1, dilated and rounded below; an oblique postmedial band from vein 10 to below vein 3 near termen; a small spot above tornus; a white dot below base of costa; a few subbasal silvery-blue scales below costa, some marks on the antemedial black band, and silvery-blue marks in and below cell, on the discocellulars, a streak on vein 2 towards termen, and a point near tornus; cilia white apically, and spotted with white towards tornus. Hind wings scarlet with a black terminal band; cilia spotted white and black. Length of fore wing 24 mm. In ♀ the thorax is spotted with pure white; the abdomen is banded yellow except the penultimate urite where the band is white; anal tuft orange. Fore wings with the markings confluent except the spot above tornus, so that the black bands are very much reduced. Length of fore wing 22 mm.

D i s t r i b u t i o n : Sierra Leone; Panyama.

Heraclia limbomaculata (STRAND)

1909 *Xanthospilopteryx limbomaculata* STRAND, Societas entomologica, Stuttgart, **24**:108. – T e r r a t y p i c a : Deutsch-Ostafrika = Tanzania (Type, ♂, in Berlin Museum).

1910 *Xanthospilopteryx discosticta* HAMPSON, Proc. zool. Soc. London, **1910**:396; pl. 36, fig. 23. – Terra typica: Alala Plateau, Ndola District, Northwest Rhodesia = Zambia (Type, ♂, in British Museum).
1920 *Heraclia limbomaculata* – HAMPSON, Cat. Lepidopt. Phal. Brit. Mus., Suppl. **2**:554.
1920 *Heraclia discosticta* – HAMPSON, Cat. Lepidopt. Phal. Brit. Mus., Suppl. **2**.555.

Diagnosis: Head and thorax black; first and second palpal joints, sides of frons, vertex, collar, tegulae, meso- and metathorax spotted with white; tibiae banded with yellow; the base and extremities of tibiae and tarsi with white points; abdomen brownish yellow in ♂, with the two last urites and underside black, with sublateral and ventral series of small yellow spots; abdomen in ♀ banded orange and black dorsally, the penultimate urite with a white band. Fore wings bluish black; whitish points at base below costa and cell; markings yellow, viz., a large triangular patch extending from below costa to vein 1; an oblique band from below middle of costa to above tornus, constricted just below the cell and more or less pointed at lower extremity, a broad band from below costa beyond middle to near termen below vein 3 where its extremity is pointed and somewhat recurved; three obliquely placed small spots across the apical area; a subterminal spot in the submedial fold; oblique silvery-blue lines in and below middle of cell and on the discocellulars, cilia spotted with white. Hind wings orange-scarlet, with a fine line on the discocellulars, and a blue-black terminal band, slightly angled inwards on vein 2; cilia spotted with white. Length of fore wing 30–36 mm.

Distribution: Tanzania, Zambia.

Heraclia mozambica (MABILLE)

1890 *Eusemia mozambica* MABILLE, Bull. Soc. entomol. France, Paris, **1890**:123. – Terra typica: Zanquebar = Zanzibar (Type, ♀, in British Museum).
1891 *Xanthospilopteryx mozambica* – KIRBY, Trans. r. entomol. Soc. London, **1891**:285.

Diagnosis: Body black; palpi, sides of frons, occiput, collar, tegulae, thorax, pectus, coxae, extremities of femora and tibiae spotted with white; fore and mid tibiae spotted with orange, abdomen with dorsal orange bands, except at the extremity, lateral series of white spots, and ventral spots on terminal urites; sublateral orange streaks on basal half. Fore wings black; several white basal and subbasal points; markings orange, viz., an oblique antemedial band from below costa to just above dorsum; a bar across the cell-extremity; a large spot in the submedian interspace; a broad band from below costa beyond middle to near termen at vein 3; a spot above tornus; silvery-blue markings in middle of cell, on the discocellulars, and postmedially on veins 2 and 1, cilia white apically. Hind wings crimson-red with the base black; a large black discocellular spot; a broad black terminal band, especially broad between veins 2 and 3, cilia with white extremities apically. Length of fore wing 32 mm.

Distribution: Tanzania.

Heraclia viettei KIRIAKOFF

1973 *Heraclia viettei* KIRIAKOFF, Lambillionea, **72–73**:58; fig. 1. – Terra typica: Nole Kaba, Ethiopia (Type, ♀, in Paris Museum).

D i a g n o s i s : Head and thorax black; first and second palpal joints, prae- and postocular region, vertex, collar, tegulae and metathorax spotted with white; legs black; tibiae with orange pilosity; tarsi banded black and white; abdomen orange with narrow black bases to the segments; underside black with three series white dots; anal tuft orange. Fore wings black with yellowish-cream markings, viz., an oblique subbasal line of three conjoined spots; a small rounded cell-spot, an elongate spot below cell, a larger, oval spot on dorsum; three dots at base of costa, and a dot below the anal vein, before the band; an oblique medial band of two spots, the upper in cell, squarish, connected with the lower one by a fine line; the lower spot more oval and more oblique, an irregularly rounded spot above tornus; postmedial band continuous, faintly incurved distally, with a small proximal notch at the base of vein 4; some dark metallic blue markings, viz., a few scales at base, a spot in cell between the bands, and an arched discocellular mark; cilia white apically. Hind wings reddish-orange, more reddish costally and on the anal region; a black terminal band, extending to middle of costa, narrowed on vein 4, slightly angled inwards on vein 2, and bearing in interspace I a small triangular orange spot; upper discocellular with a small oval black spot; cilia broadly white apically. Length of fore wing 24,5 mm.

D i s t r i b u t i o n : Ethiopia.

Genus **Pentelia gen. nov.**

(fig. 6)

S p e c i e s t y p i c a : *Eusemia pentelia* DRUCE (by tautonomy).

D i a g n o s i s : Proboscis fully developed; antennae dilated towards the extremity, palpi upturned, with the first and second joints clothed with rather long hair in front, the third long, naked; frons with a truncate conical projection with a raised rim at the extremity; hind tibiae clothed with rough hair. Nervation: in fore wings, vein 2 arising at about three quarters of cell; 3, 4 from about a point; 5 slightly separated; discocellulars incurved; 6 from the upper cell-angle; areole narrow; 7 and the stalk of 8 + 9 from its extremity; 10 from its upper angle. In hind wings, vein 2 arising at three-quarters of cell; 3, 4 and 6, 7 from the cell-angles; discocellulars incurved; 5 obsolescent, from middle thereof.

Male genitalia: Uncus slightly shorter than costa, compressed, regularly arched in the proximal half, then very faintly so. Tubus analis slightly longer than uncus, narrow, well sclerified. Vinculum produced at middle in a triangular lobe, with a tuft of short stiff hair. Valva broad, upcurved, broadening distally; costa more or less regularly but very moderately incurved; apex rounded; sacculus fairly regularly outcurved; fold broad, narrowing distally, not reaching the middle of valva, and ending in a long harpe, hooked terminally only. Aedeagus longer than one-and-a-half times the costa, very robust, straight, except in the terminal third where it is slightly bent and partly covered with half-microscopic spinules; fultura inferior well developed; proximal margin outcurved at middle, and with the angles produced; sides strongly incurved; distal margin with the angles acutely produced. Saccus semi-elliptical.

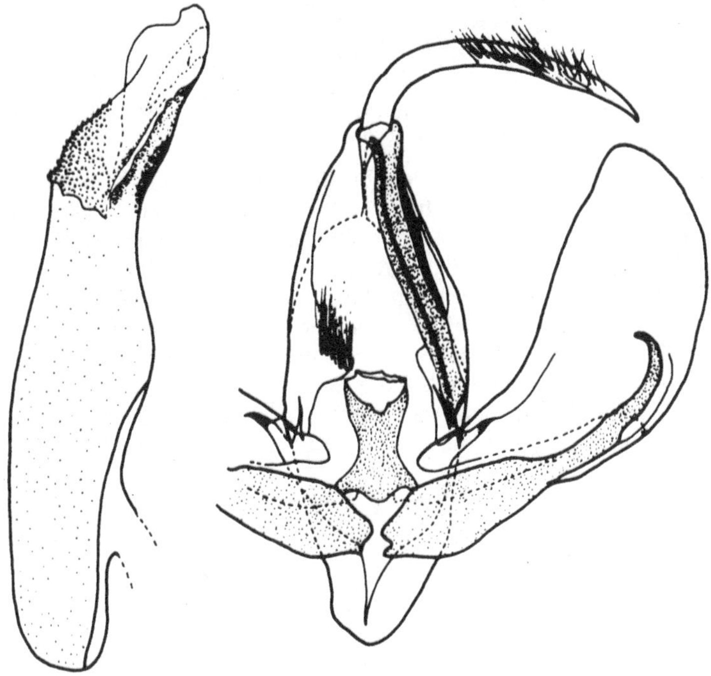

Fig. 6. *Pentelia pentelia* (DRUCE). B. M. Noct. 8279. Typus

This new genus differs from the nearly related genera *Heraclia* and *Rothia* (where *E. pentelia* is currently placed) in details of the male genitalic structures.

Distribution: East Africa.

Pentelia pentelia (DRUCE)

1887 *Eusemia pentelia* DRUCE, Proc. zool. Soc. London, **1887**:668. – Terra typica: East Africa, Delagoa Bay (Type, ♂, in British Museum).

Diagnosis: Head and thorax black; sides of palpi and of frons, occiput, collar, tegulae and thorax above spotted with yellowish-white; coxae, femora and tibiae with orange pilosity; tarsi ringed with white; abdomen black with orange rings, the three last segments with bands reduced to spots on the underside; anal tuft orange. Fore wings black; markings yellowish, viz., two points at the base of costa; antemedial spots in cell and above vein 1; a bar across cell; a rounded patch on the dorsal area; a postmedial patch between veins 2 and 8; a spot above tornus; silvery-blue bars in middle of cell and on the discocellulars. Hind wings orange red, with a black terminal and band excised at middle and not reaching middle of wing on costa. Length of fore wing 24 mm.

Distribution: East Africa, Delagoa Bay.

Genus **Rothia** WESTWOOD
(fig. 7)

1877 *Rothia* WESTWOOD, Trans. Linn. Soc. London, Zool., (2) **1**:204. – Species typica: *Agarista pales* GUÉRIN (by original designation).

Diagnosis: Proboscis fully developed; antennae distinctly dilated towards the extremity; palpi upturned; the first and the second joints with rather long hair in front; the third joint well developed, naked, somewhat porrect; frons with a truncate conical priminence, with a slight raised rim at the extremity; hind tibiae more or less strongly clothed with rough hair above. Nervation: in fore wings, vein 2 arising at about three-quarters of cell; 3 and 4 from a point; 5 from the lower angle of cell; discocellulars angled inwards; 6 from the upper cell-angle; areole narrow; 7 and the stalk of 8 + 9 from its extremity; 10 from the upper angle thereof. In hind wings, vein 2 arising at three-quarters of cell; 3, 4 and 6, 7 from the cell-angles; discocellulars faintly incurved; 5 obsolescent, from the middle thereof.

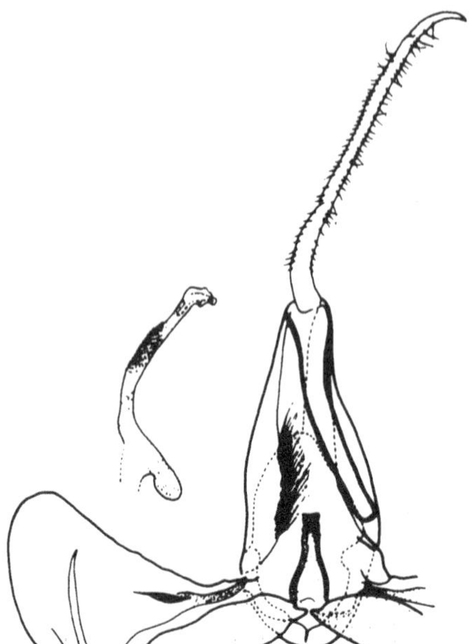

Fig. 7. *Rothia panganica* KARSCH. B. M. Noct. 8281

Male genitalia: Uncus very long, as long or longer than valva, compressed, arched. Tubus analis elongate, sometimes as long as uncus. Vinculum bearing a bunch of stiff hair. Valva shaped like an axe, with a rather broad fold of the sacculus, and a short, arched harpe. Aedeagus rather short, bearing in most species a long and slender terminal process, sometimes bent at a right angle at middle; fultura inferior variable. Saccus short, rounded.

This genus, rich in species in Madagascar, is represented on the African continent by a single species.

D i s t r i b u t i o n : Madagascar; East and Southeast Africa.

Rothia panganica KARSCH

1892 *Rothia panganica* KARSCH, Entomol. Nachr.Bl., Wien, **24**:330. – T e r r a t y p i c a : Deutsch-Ostafrika = Tanzania (Type, ♂, in British Museum).

1974 *Rothia rhaeoides* KIRIAKOFF, Bull. Ann. Soc. r. entomol. Bel., Bruxelles, **110**:150, fig. 4. – T e r r a t y p i c a : Malawi: Mount Mlanje (Type, ♂, in British Museum).

D i a g n o s i s : Body and wings black; palpi, head, collar and tegulae spotted with creamy white; fore tibiae and femora banded with orange; a small orange crest on the first abdominal tergite; tip of abdomen orange. Fore wings with small creamy white spots at base, middle and end of cell above the media; a squarish spot beyond upper angle of cell, and another, rounded, beyond the lower cell-angle, spots sometimes conjoined; silvery blue spots between the white cell-spots, on the discocellulars, in interspace I before one-third and before two-thirds, and below vein 1 near the tornus; cilia white apically. Hind wings with a creamy-white spot at termen below the media, and sometimes below the anal angle; cilia white apically and near the anal angle. Length of fore wing 19–19,5 mm.

R e m a r k s : HAMPSON (Cat. Lepidopt. Phal. Brit. Mus., Suppl. 2:569, 1920) described an aberration 1, without cell-spots and with the spots in hind wings conjoined; such specimens, restricted to Malawi, Mt. Mlanje (of which there is a series in the British Museum), correspond to *R. rhaeoides* KIRIAKOFF, and possibly represent a local form and even a distinct species; the male genitalia of both forms, however, differ in no way, reason why they are considered in the present revision as conspecific, pending further information.

D i s t r i b u t i o n : Tanzania, Malawi.

Genus Chaetostephana JORDAN

(fig. 8)

1913 *Chaetostephana* JORDAN in SEITZ, Großschmetterl Erde, **15**:20. – S p e c i e s t y p i c a : *Metagarista rendalli* ROTHSCHILD (by original designation).

D i a g n o s i s : Proboscis fully developed. Antennae dilated distally in both sexes. Frons in both sexes broad, strongly convex, with a weakly developed, small prominence without raised rim. Palpi oblique; second and third joints clothed with moderately long hair, the third somewhat rough and rather robust, not slender as in *Hespagarista*. Hind tibiae clothed with rough hair. Abdomen with a slight dorsal crest at the base, and without long anal hair in male, as in *Hespagarista*. Nervation: In fore wings, vein 2 from slightly less than three-quarters of cell; 3, 4, 5 rather well separated; 6 from the upper cell-angle; areole long and narrow, with veins 7 and 8 + 9 stalked from its extremity; 10 from upperside of the areole. In hind wings, vein 2 from three-quarters of cell; 3, 4 and 6, 7 from the cell-angles; discocellulars nearly straight; 5 obsolescent, from middle of the discocellulars.

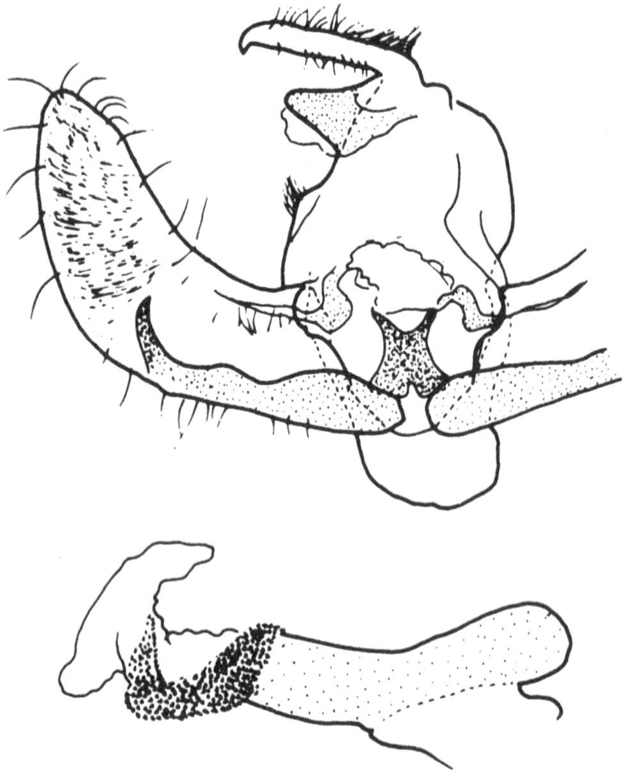

Fig. 8. *Chaetostephana rendalli* (ROTHSCHILD). B. M. Noct. 8614. Typus

Male genitalia: Uncus about as long as half of the costa, compressed, bent at a straight angle basally, then nearly straight, with a short, curved terminal hook. Tubus analis short and broad, well sclerified. Valva moderately broad, well upcurved, with the costa and sacculus nearly parallel; apex rounded; fold of sacculus narrow, reaching to middle of valva and ending in a small, curved harpe. Aedeagus slightly longer than three-quarters of the sacculus, robust, with the terminal portion densely covered with minute spinelike cornuti; vesica also bearing a patch of cornuti; fultura inferior broad basally, notched at middle of the proximal margin, with the sides strongly incurved, and the distal margin with narrow and produced angles. Saccus squarish with rounded angles.

Distribution: West, Central and East Africa.

Key to the species

1 Distal abdominal segments from the sixth black . . . . **Ch. rendalli**, p. 43
1′ Only the last abdominal segment black . . . . **Ch. inclusa**, p. 43

Chaetostephana rendalli (ROTHSCHILD)

1896 *Metagarista rendalli* ROTHSCHILD, Novit. Zool., **3**:97. – Terra typica: Zomba, British Central Africa (Type, ♂, in British Museum).
1901 *Hespagarista rendalli* – HAMPSON, Cat. Lepidopt. Phal. Brit. Mus., **3**:631, fig. 277.
1913 *Chaetostephana rendalli* – JORDAN in SEITZ, Großschmetterl. Erde, **15**:21.

Diagnosis: Head and thorax black, collar orange; basal abdominal crest orange, first and second urites black banded orange, third, fourth and fifth orange, rest black; underside black spotted with orange. Fore wings black; an oblique, roughly triangular orange antemedial patch from below the costa to vein 1 or below; a rounded discocellular spot, an oblique more or less elliptical patch beyond the cell from below the costa to vein 2; silvery blue markings, viz.: a spot below the base of the costa, a bar across the cell-middle, a discocellular bar, an oblique streak below vein 2 with a spot below it, and two conjoined subapical spots. Hind wings orange, with a broad black terminal band. Length of fore wing 19–23 mm.

Distribution: Central East Africa; Angola.

Chaetostephana rendalli rendalli (ROTHSCHILD)

1913 *Chaetostephana rendalli* ssp. *rendalli* JORDAN in SEITZ, Großschmetterl. Erde, **15**:21; pl. 2d.

Diagnosis: As above.

Distribution: Tanzania, Zaire (Shaba).

Chaetostephana rendalli angolana JORDAN

1913 *Chaetostephana rendalli* ssp. *angolana* JORDAN in SEITZ, Großschmetterl. Erde, **15**:21. – Terra typica: Bihé, Angola (Types, ♂, ♀, in British Museum).

Diagnosis: Yellow markings slightly enlarged and paler; marginal band on hind wings slightly narrower.

Distribution: Angola.

Chaetostephana inclusa (KARSCH)

1895 *Pristoceraea inclusa* KARSCH, Entomol. Nachr.Bl., Wien, **21**:357; pl. 2, fig. 5. – Terra typica: Guinea (Type, ♀, in Berlin Museum).
1901 *Hespagarista inclusa* – HAMPSON, Cat. Lepidopt. Phal. Brit. Mus., **3**:631; pl. 53, fig. 3.
1913 *Chaetostephana inclusa* – JORDAN in SEITZ, Großschmetterl. Erde, **15**:21; pl. 2e.

Diagnosis: Female. Head and thorax black; collar orange; abdomen orange, with slight segmental lines of orange, extreme tip black; underside black spotted with orange. Fore wings black; a large orange subbasal patch from below the costa to vein 1, expanding in the submedial interspace and extending to just beyond the middle, and just below vein 1, an elliptical spot at the cell-tip; a large postmedial spot from below the costa to below vein 2, rounded at the extremities, some blue markings, viz.: a basal spot, a bar across the cell, a discocellular bar, and postmedial spots on veins 1 and 2. Hind wings orange, with a rather narrow black terminal band, with the proximal edge wavy. Length of fore wing 23 mm.

Distribution: West Africa: Guinea; (?) Angola (Rothschild collection); Shaba (?) (Tervuren Museum).

Genus **Hespagarista** WALKER

(fig. 9)

1854 *Hespagarista* WALKER, List Lepidopt. Het. Brit. Mus., **1**:13. – Species typica: *Agarista echione* BOISDUVAL (by monotypy).
1874 *Amazela* BOISDUVAL, Rev. Mag. Zool., (2) **3**:67. – Species typica: *Agarista calisto* BOISDUVAL (by subsequent designation by HAMPSON, 1910).

Diagnosis: Proboscis fully developed. Antennae distinctly dilated towards the extremity in both sexes. Palpi oblique, clothed with long hair, the terminal joint short; frons with a truncate conical priminence with a raised terminal rim. Thorax, femora and abdomen clothed with rough hair; tibiae without spines, clothed with moderately rough hair; first joint of hind tarsi without long hair. Anal tuft very long in male. Nervation: in fore wings, vein 2 given off rather distally, at about five-sixths of the cell; 3, 4, 5 separate and equidistant; discocellulars faintly incurved; 6 from the upper cell-angle; areole long and narrow; 7 and the stalk of 8 + 9 from its extremity; 10 from the upper margin of areole, just before its extremity. In hind wings vein 2 given off at five-sevenths of the cell; 3, 4 and 6, 7 from the cell-angles; discocellulars incurved; 5 obsolescent, from middle of the discocellulars.

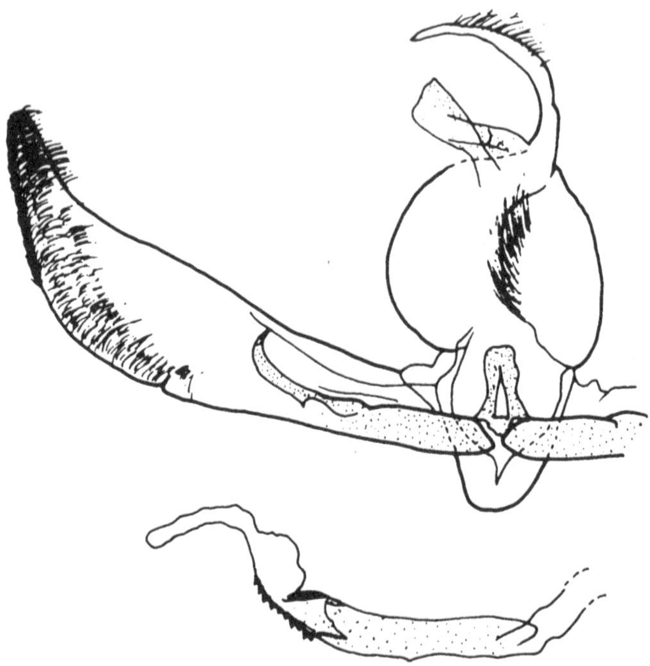

Fig. 9. *Hespagarista caudata* (DEWITZ). B. M. Noct. 8613

Male genitalia: Uncus slender, compressed, faintly arched, about one-third of valva in length, ending in a slender hook. Tubus analis short and rather broad, well sclerified. Fold of vinculum broad, with a hair tuft in middle portion. Valva relatively very elongate, narrow at base, then broadening, especially costally, but narrowing again in the distal third to end in an acute apex; costa incurved basally and distally, and outcurved medially; sacculus regularly outcurved; terminal fifth of sacculus with a fringe of short stiff spinelike hair; fold very narrow, only reaching to about one-quarter of sacculus, and ending in a very small, slender, arched harpe. Aedeagus slightly shorter than one-half of costa, rather slender, faintly arched, with a terminal row of denticulations; fultura inferior small, narrow, angulate proximally, with the angles slightly produced, and deeply cleft at middle. Saccus roughly semi-circular, pointed.

Distribution: South and southern Central Africa.

Key to the species

1 Head orange above; hind wings with several whitish spots, varying in size *H. echione*, p. 45
1′ Head black above; hind wings with a single very large white patch . . . . *H. caudata*, p. 46

Hespagarista echione (Boisduval)

1847 *Agarista echione* BOISDUVAL, Voyage Deleg., **2**:595. – Terra typica: South Africa (Type, ♂, in British Museum).
1854 *Hespagarista echione* – WALKER, List Lepidopt. Het. Brit. Mus., **1**:13.
1854 *Hespagarista interlecta* WALKER, List Lepidopt. Het. Brit. Mus., **1**:14. – Terra typica: Zomba, British Central Africa = Malawi (Type, ♂, in British Museum).
1874 *Amazela echione* – BOISDUVAL, Rev. Mag. Zool., (2) **3**:67.
1877 *Hespagarista interjecta* WESTWOOD, (ex errore pro *interlecta*), Trans. Linn. Soc. London, (2) **1**:203.
1890 *Eusemia novemmaculata* MABILLE, Ann. Soc. entomol. France, (6) **10**:54. – Terra typica: "Caffraria" (Type, ♂, in British Museum).
1892 *Anaphela novemmaculata* – KIRBY, Syn. Cat. Lepidopt. Het.:25.
1896 *Hespagarista echione* var. *funebris* ROTHSCHILD, Novit. Zool., **3**:44. – Terra typica: Tanzania (Type, ♀, in British Museum).
1913 *Hespagarista echione* aberr. *spilota* JORDAN in SEITZ, Großschmetterl. Erde, **15**:21; pl. 2e (as *funebris*). – Terra typica designata (hoc loco): Tanzania (Type, ♀, in British Museum).

Diagnosis: Body and wings black. Head orange above; first and second palpal joints with white points at the extremities; thorax orange above; tibiae with white points; abdomen with lateral and sublateral series of white points. Fore wings with an ochreous dot at the base, subbasal points below costa and above vein 1; a spot in middle of cell; a discocellular bar; a large rounded spot below middle of cell; large spots beyond both cell-angles; a spot above tornus; some silvery blue scales, viz.: below cell near the base; an antemedial line; an oblique bar across the cell-extremity; a postmedial line bent inwards in the costal area. Hind wings with pale ochreous patches in and below cell, and wedge-shaped spots beyond the cell between veins 2 and 6; cilia sometimes white at the extremities. The light markings sometimes are more or less reduced (♀ aberr. *spilota*) or nearly obsolescent (♀ aberr. *funebris*). Length of fore wing 22–25 mm.

D i s t r i b u t i o n : Natal, Malawi, Moçambique, Tanzania, Burundi. One ♀ of the form *spilota* in Paris Museum labelled Dahomey 1907 (Eug. Poisson). Dark ♀ – forms recorded chiefly from the northern parts of the area.

Hespagarista caudata (DEWITZ)

1879 *Agarista caudata* DEWITZ, Mitt. münchner entomol. Ges., 3:30; pl. 1, fig. 3 and 3a. – T e r r a t y p i c a : Guinea (Type, ♂, in Berlin Museum?).
1892 *Hespagarista caudata* – KIRBY, Cat. Lepidopt. Het.:21.
1901 *Hespagarista caudata* – HAMPSON, Cat. Lepidopt. Phal. Brit. Mus., 3:630.

D i a g n o s i s : Body and wings black; palpi and tibiae with white points; prothorax above and tegulae orange; abdomen with lateral and sublateral series of white points. Fore wings with pale ochreous markings, viz.: basal and subbasal points, antemedial strigae above and below vein 1; an elliptical cell-spot, with a point above it on costa; a square spot below middle of cell, conjoined to a spot below vein 1; a discocellular spot; a postmedial costal spot; a series of four or five subterminal elongate spots; some silvery-blue markings, viz.: an antemedial line, a spot in cell-extremity, and a sinuous postmedial line; cilia white apically. Hind wings with a broad pale ochreous band from below middle of costa to near anal margin above anal angle, with dark veins on it; mostly a fine black discocellular lunule. Length of fore wing 19–24 mm in ♂, 23–28 mm in ♀.

D i s t r i b u t i o n : Guinea; Angola (fide JORDAN 1900), "Kongo" (fide STRAND 1900), Zambia, Shaba.

H e s p a g a r i s t a c a u d a t a c a u d a t a (DEWITZ) comb. nov.

D i s t r i b u t i o n : Guinea, Angola to Shaba and Zambia.

H e s p a g a r i s t a c a u d a t a e b u r n e a JORDAN

1915 *Hespagarista caudata* ssp. *eburnea* JORDAN, Novit. Zool., 22:274; fig. 1. – T e r r a t y p i c a : Manow, Tanzania (Type, ♂, in British Museum).

D i a g n o s i s : Markings paler, yellowish white, somewhat more extensive.

D i s t r i b u t i o n : Tanzania.

Genus Syfanoidea BARTEL

(fig. 10)

1903 *Syfanoidea* BARTEL, Verh. zool.-bot. Ges. Wien, 53:125. – S p e c i e s t y p i c a : *Syfanoidea schencki* BARTEL (by original designation).

D i a g n o s i s : Proboscis fully developed. Antennae distinctly dilated towards the extremity. Palpi short, not reaching beyond frons, clothed with rather long hair; terminal joint short, somewhat rough scaled. Frons broad in both sexes, with a well developed conical truncate prominence, with a raised rim at the extremity, and with a flat process beneath it, ending in an upturned spine. Fore tibiae without terminal spine in ♀; tibiae fringed with long hair, and without spines. Abdomen with a strong dorsal basal crest; anal tuft short in ♀. Fore wings rather broad, with the apex rounded. Nervation: in fore wings, veins 3, 4, 5 much approximated; discocellulars strongly incurved; 6 arising from the lower side of the areole near its extremity; 7 and the stalk of 8 + 9 on a common stalk from the extremity of the areole; 10 from the upper side of the areole, near its extremity. In hind wings, veins 3 and 4 from the lower cell-angle; 6 and 7 very short stalked.

Male genitalia: Uncus about two-fifths of costa in length, narrow, arched, with beak-like extremity. Tubus analis short and broad, sclerified ventrally. Vinculum moderately broad, with a fringe of moderately long, stiff hair. Valva banana-shaped, with the costa and sacculus nearly parallel, the former incurved, the latter outcurved; apex blunt; a row of stiff curved setae on apex and along the distal fifth of the sacculus; fold of sacculus rather narrow, reaching slightly beyond middle of the valva, and ending in a short, sickle-shaped harpe. Aedeagus about five-sevenths of costa in length, robust, nearly straight, bearing terminally a lateral row of a few minute cornuti; fultura inferior very deeply excised at middle of the proximal margin, slightly narrowing distally, and with the terminal margin faintly produced at the angles. Saccus short, rounded.

Distribution: South Africa.

Syfanoidea schencki BARTEL

1903 *Syfanoidea schencki* BARTEL, Verh. zool.-bot. Ges. Wien, **53**:126–127. – Terra typica: Natal (Type, ♀, in Museum Berlin).
1906 *Aegocera leighi* JORDAN, Novit. Zool., **13**:161. – Terra typica: Durban, Natal (Type, ♂, in British Museum).
1920 *Aegocera schenci* HAMPSON (ex errore pro *schencki*), Cat. Lepidopt. Phal. Brit. Mus., Suppl. **2**:577; pl. 71, fig. 17.

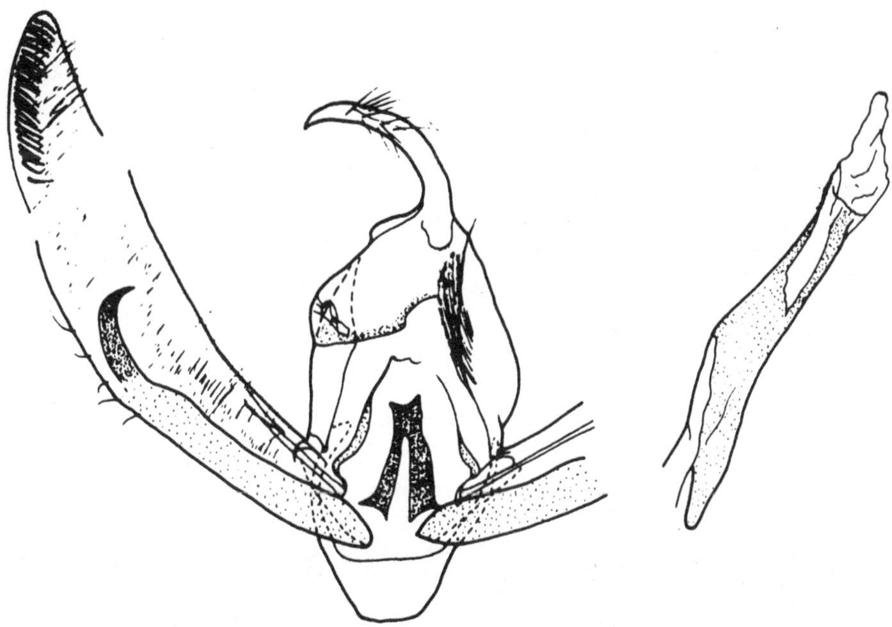

Fig. 10. *Syfanoidea schencki* (BARTEL). B. M. Noct. 8483

D i a g n o s i s : Head yellowish white; frons with some red scales; antennae black towards the extremity; palpi and sides of frons red; extremity of the third palpal joint black; vertex with a black dot; thorax red mixed yellowish white scales; pectus grey mixed rufous brown; legs orange; fore coxae red; fore and mid tibiae with two black spots; tarsi black spots, and a lateral series of points; underside rufous white. Fore wings rufous brown, sprinkled with grey; a yellowish white fascia from base of costa along the media to beyond base of vein 2, expanding into triangular patches at the cell-middle to the submedial fold at base of vein 2; cell- and discocellular marks obsolescent; an oblique yellowish white bar from costa beyond the middle to the submedial fold; terminal area much paler, yellowish white, tinged and sprinkled with red; a terminal series of small red lunules, edged proximally with yellowish-white lunules. Hind wings orange; a small black discocellular spot; a submarginal series of black lunules, those in interspaces II and III small and round; spot in the submedial interspace large. Length of fore wing 16 mm.

D i s t r i b u t i o n : Republic of South Africa (East Cape Province, Natal).

Genus Paraegocera HAMPSON

(fig. 11)

1901 *Paraegocera* HAMPSON, Cat. Lepidopt. Phal. Brit. Mus., **3**:607; fig. 265.– S p e c i e s t y p i c a : *Aegocera confluens* WEYMER (by original designation).

D i a g n o s i s : Proboscis fully developed. Palpi porrect, the second joint fringed with long hair below, the third naked, moderate, porrect. Antennae dilated before the extremity. Frons with a short rather beak-like prominence. Tibiae fringed with long hair above. Abdomen with dorsal crests on basal segments; male with long lateral tufts of hair from the base. Nervation: Fore wings with vein 2 from slightly beyond two-thirds of cell; 3, 4, 5 separated, distance between 3 and 4 about twice that between 4 and 5; discocellulars incurved; areole narrow; 7 and stalk of 8 + 9 from the lower angle of areole; 10 from the upper angle thereof. Hind wings with vein 2 from five-sixths of cell; 3, 4 and 6, 7 from the cell-angles; discocellulars faintly incurved.

Male genitalia: Uncus moderate, slightly longer than one half of costa, compressed, bent only at the base, then nearly straight, with a small terminal hook. Tubus analis rather narrow, sclerified ventrally. Valva banana-shaped, but slighly broader in the terminal half; fold of sacculus broad basally, about three-quarters of valva in width, then narrowing and reaching the middle of the sacculus; harpe small, sickle-shaped. Aedeagus slightly shorter than costa, moderately robust, faintly arched, cut squarely at the distal end; fultura inferior rather small, three-pronged proximally, cut straight distally; fultura superior membraneous. Saccus short semi-elliptical.

D i s t r i b u t i o n : East Africa.

Paraegocera confluens (WEYMER)

1892 *Aegocera confluens* WEYMER, Stettin. entomol. Ztg., **53**:100. – T e r r a t y p i c a : Zanzibar (Type, ♀, in British Museum).

1895 *Aegocera darocana* DRUCE, Ann. Mag. nat. Hist., London, (6) **15**:42. – T e r r a t y p i c a : Lindi, Tanzania (Type, ♂, in British Museum).

1901 *Paraegocera confluens* – HAMPSON, Cat. Lepidopt. Phal. Brit. Mus., **3**:607.
1913 *Schausia confluens* – JORDAN in SEITZ, Großschmetterl. Erde, **13**:16; pl. 4b.

Diagnosis: Head and thorax deep chocolate brown mixed with some yellowish white; mid and hind tibiae mostly orange; abdomen orange, with the tufts at the base chocolate brown and white, and the anal tuft and the underside towards the extremity blackish. Fore wings deep chocolate brown; a large yellowish-white patch in proximal half formed by the fusion of a patch at base of costa, the antemedial patch from middle of cell to vein 1, spots in and below middle of cell, and a

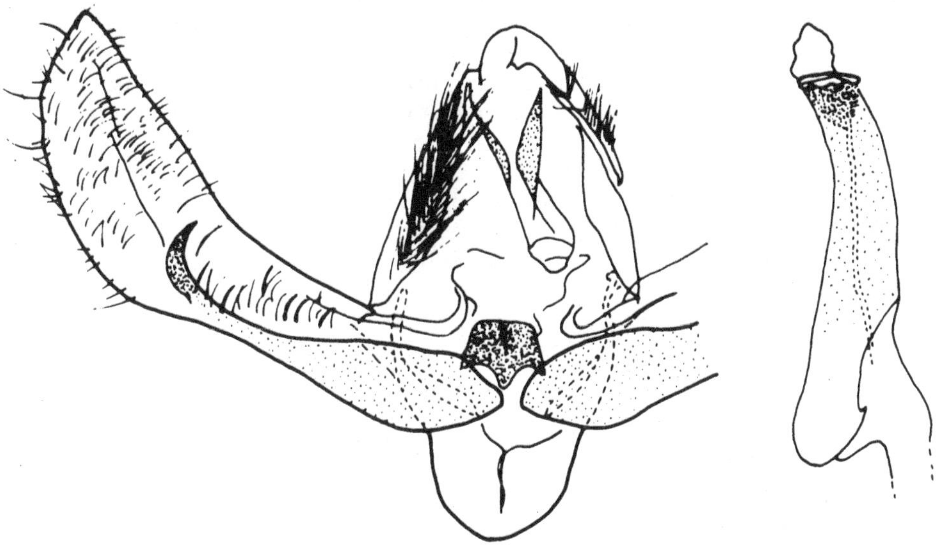

Fig. 11. *Paraegocera confluens* (WEYMER). B. M. Noct. 8427. Typus

spot from below the cell-end to dorsum; additional yellowish-white markings as follows: a spot in upper end of cell; two conjoined square spots beyond cell on costa and between veins 3 and 5; and a terminal band narrowing to apex and tornus, with the proximal outline irregular; silvery blue markings, viz.: antemedial line from costa to patch in cell; a discocellular bar; and a diffused subterminal line excurved from costa to vein 2, then incurved and expanding into a patch; a fine black terminal line; cilia white spotted with black. Hind wings orange with a black terminal band narrowing from costa to a point on vein 1. Length of fore wing 16 mm.

Distribution: Tanzania; Uganda, Kenya.

Genus **Metagarista** WALKER

(fig. 12)

1854 *Metagarista* WALKER, List Lepidopt. Het. Brit. Mus., 1:60. – Species typica: *Metagarista triphaenoides* WALKER (by monotypy).

Diagnosis: Proboscis fully developed. Palpi obliquely upturned, with the second joint fringed with long rough hair in front, and the third short and hairy.

Antennae distinctly dilated at the extremity in male, less so in female. Frons narrower than the eye is high in frontal view, narrower in female, frontal cone short, broad, truncate, with a raised rim at the extremity. Tibiae fringed with hair; hind tarsi in male with the first two joints fringed with long hair; hind legs in female with appressed scales. Abdomen with dorsal crests on basal segments; in male, with lateral protrusible tufts at the base.

Male genitalia: Uncus about one-third of costa in length, narrow, especially distally, pointed and slightly hooked terminally. Tubus analis short, broad, sclerified laterally. Vincula very broadly fused and with narrow folds bearing distally a tuft of hair. Valva rather narrow at the base, then broadening; costa more or less straight, faintly incurved terminally; sacculus distinctly convex at middle, more or less straight in the proximal half, outcurved distally; apex rounded, bearing a short series of stiff hairs; fold of the sacculus narrow, not reaching the middle of the valva, and ending in a rather short, scythe-shaped harpe, distinctly broader in its middle section. Aedeagus slightly longer than three-quarters of costa, rather slender, nearly straight, bearing terminally lateral patches of minutest cornuit; vesica with a small lateral lobe; fultura inferior roughly triangular, with the sides slightly incurved subterminally; distal margin excised at middle; a small oval patch at the middle of the proximal margin; fultura superior narrowly sclerified laterally. Saccus semi-elliptical.

Distribution: West and Central Africa.

Key to the species

1 Abdomen black above, banded with white; fore wings dark reddish brown, with an obsolete whitish postdiscal patch; hind wings with the terminal band complete . . . . . *M. triphaenoides,* p. 50

1′ Abdomen black above, banded with orange; fore wings greyish yellowish brown, with fine black transverse lines; hind wings with a terminal series of black spots *M. maenas,* p. 51

Metagarista triphaenoides WALKER

1854 *Metagarista triphaenoides* WALKER, List, Lepidopt. Het. Brit. Mus., 1:61. – Terra typica: Gold Coast (Type, ♂, in British Museum).

Diagnosis: Head rufous; thorax ochre-grey; collar and tegulae edged with rufous; pectus orange; fore legs rufous; tarsi black; abdomen black with white bands, base orange, dorsal crests rufous; underside orange with black segmental lines. Fore wings brownish, more or less tinged with olive or reddish, and very thickly sprinkled with chocolate brown; antemedial line sinuous, silvery and chocolate, angled at the cell; cell- and discocellular markings olive, edged with chocolate and centred with silvery; postmedial line chocolate, excurved to vein 3, then incurved, and with some silver distally; a more or less undistinct whitish postmedial patch; some chocolate suffusion on the apical and dorsal areas; cilia chocolate. Hind wings reddish orange, with a blackish brown terminal band of nearly uniform breadth, and some white on the termen towards the anal angle. Length of fore wing 30 mm.

Distribution: Gold Coast, South Nigeria, Sierra Leone, Ivory Coast, Cameroon, Senegambia, to Uganda; great part of Zaire; Burundi.

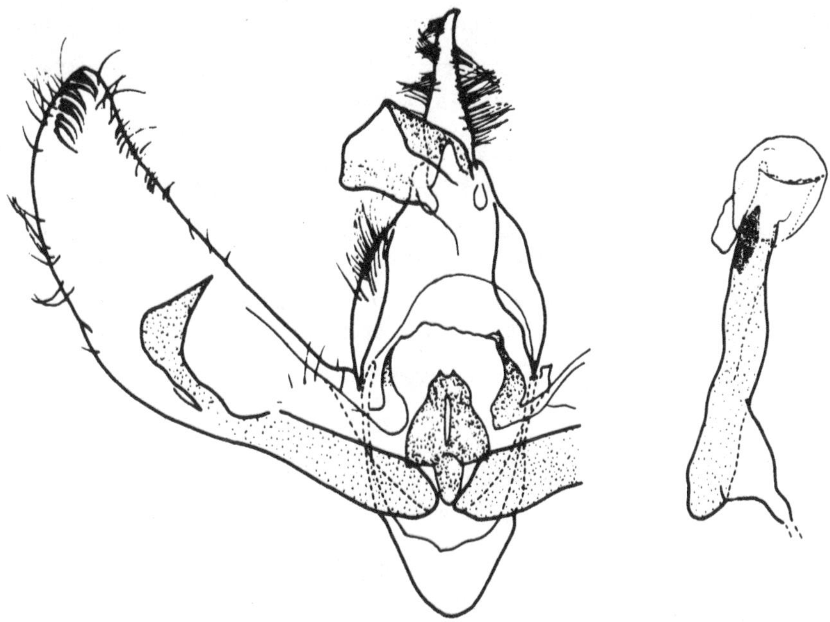

Fig. 12. *Metagarista triphaenoides* WALKER. B. M. Noct. 8494. Typus

Metagarista maenas (HERRICH-SCHÄFFER)

1853 *Aegocera maenas* HERRICH-SCHÄFFER, Außereurop. Schmetterl.: figs. 20 and 21. – Terra typica: West Africa (Type, ♂, in British Museum).
1892 *Metagarista maenas* – KIRBY, Syn. Cat. Lepidopt. Het.:41.

Diagnosis: Head orange; spot on the second and the third palpal joints black; head with a dark median stripe; thorax olive grey; collar and tegulae edged with rufous; femora striped with rufous; pectus and legs orange; fore femora above, and tarsi black; abdomen black with orange bands, and the underside orange. Fore wings olive grey; costal and dorsal areas to the postmedial line browner and sprinkled with chocolate brown; a sinuous antemedial line, chocolate brown, from costa to below cell; spots in cell, on the discocellulars, and below cell, edged and centred with chocolate brown, and with a leaden line at middle; veins beyond the cell streaked with chocolate, as are the interspaces in the terminal area; a double slightly waved postmedial line from costa to interspace ii, excurved, then incurved-outer line with some leaden grey on its proximal area; a fine waved chocolate terminal line; cilia chocolate at the extremities. Hind wings rufous orange, with a series of red brown terminal spots. Length of fore wing 26–29 mm.

Distribution: West Africa to Zaire and Uganda.

Metagarista maenas maenas (HERRICH-SCHÄFFER)

1913 *Metagarista maenas* ssp. *maenas* – JORDAN in SEITZ, Großschmetterl. Erde, **15**:15; pl. 2g.

D i a g n o s i s : Redbrown terminal spots on hind wings more or less confluent.

D i s t r i b u t i o n : West Africa (Senegambia, Dahomey, Gabon, Sierra Leone, Gold Coast, Nigeria, Congo [Brazzaville], Cameroon).

Metagarista maenas disjuncta JORDAN

1913 *Metagarista maenas* ssp. *disjuncta* JORDAN in SEITZ, Großschmetterl. Erde, **15**:15. – T e r r a t y p i c a : Bopoto, Congo (Type, ♀, in British Museum).

D i a g n o s i s : Reddish-brown terminal spots on hind wings much narrower than the interspaces, and connected by a fine obsolescent line.

D i s t r i b u t i o n : Most of the Congo-basin; Uganda.

Genus **Polacanthopoda** HAMPSON

(fig. 13)

1901 *Polacanthopoda* HAMPSON, Cat. Lepidopt. Phal. Brit. Mus., **3**:573. – S p e c i e s t y p i c a : *Hespagarista tigrina* DRUCE (by original designation).

D i a g n o s i s : Proboscis fully developed; antennae distinctly dilated towards the extremity; palpi upturned, with the second joint broadly scaled in front, and the third short, with rough scales in front; frons with rounded prominence and narrowed above; mid and hind tibiae spined, the tibiae fringed with long hair on the outer side; abdomen with a dorsal tuft on the basal urite. Venation: in fore wings, vein 2 from about three-quarters of cell; 3, 4, 5 well separated and equidistant; discocellulars incurved; 6 from upper angle of cell; areole narrow; 7 and stalk of 8 + 9 from the extremity of areole; 10 from the upper margin of areole, near the extremity. Hind wings with vein 2 from three-quarters of cell; 3, 4 and 6, 7 from the cell-angles; discocellulars incurved; 5 obsolescent, from middle of discocellulars. Male with large lateral tufts of hair at the base of hind wings.

Male genitalia: Uncus about one-half of costa in length, narrow, faintly arched, rounded terminally. Vinculum with the fold narrow, bearing a row of short hairs. Valva rather broad; costa and sacculus nearly parallel; apex slightly upturned, rounded; termen oblique; fold of sacculus rather broad, only reaching to two-fifths of valva, followed by a long and robust harpe, straight for three-quarters of its length, then arched. Aedeagus distinctly longer than valva, moderately robust, nearly straight, expanded basally; fultura inferior narrow and elongate, with the sides nearly parallel. Saccus semi-circular.

D i s t r i b u t i o n : Most of West and Central Africa.

Key to the species

1 Fore wings with a triangular creamy-white patch from base to near discocellulars; hind wings chocolate-brown with a large orange medial patch . . . . ***P. naveli***, p. 56
1' Fore wings with the markings different, hind wings with only a dark terminal band . . . 2
2 Fore wings black with chocolate-red terminal band; basal area with two yellow oblique bands . . . . ***P. tigrina***, p. 53
2' Fore wings dark or chocolate-brown; basal area without yellow oblique bands 3

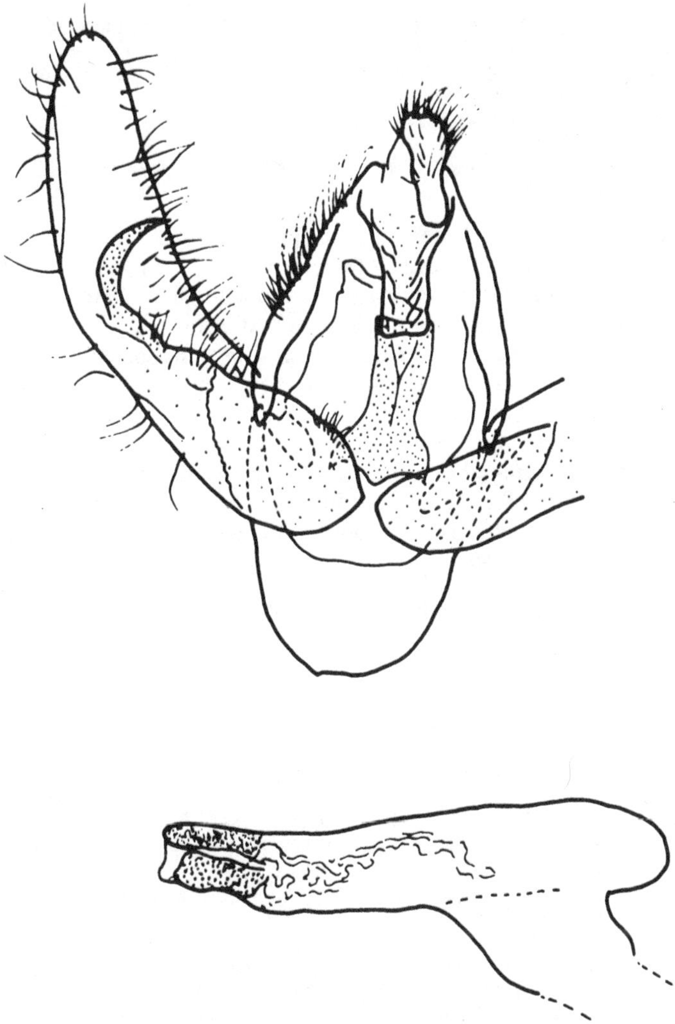

Fig. 13. *Polacanthopoda tigrina* (DRUCE). B. M. Noct. 8562. Typus

3 Fore wings with a postmedial series of whitish spots; hind wings orange-red, without discocellular mark . . . . **P. humphreyi**, p. 54
3' Fore wings with a large roundish preapical yellow-buff patch; hind wings ochreous yellow, with a black discocellular mark . . . . **P. anthina**, p. 55

Polacanthopoda tigrina (DRUCE)

1882 *Hespagarista tigrina* DRUCE, Proc. zool. Soc. London, **1882**:788; pl. 9, fig. 4. – Terra typica: Old Calabar (Type, ♂, in British Museum).
1891 *Xanthospilopteryx* (?) *tigrina* – KIRBY, Trans. r. entomol. Soc. London, **1891**:282.
1892 *Metagarista* (?) *tigrina* – HAMPSON, Cat. Lepidopt. Phal. Brit. Mus., **3**:573; fig. 282.

1913 *Aegocera tigrina* – JORDAN in SEITZ, Großschmetterl. Erde, **15**:17; pl. 4d.
1913 *Aegocera tigrina* forma *castaneimargo* JORDAN in SEITZ, Großschmetterl. Erde **15**:17. – (Type, in British Museum).
1913 *Aegocera tigrina* forma *nigrimargo* JORDAN in SEITZ, Großschmetterl. Erde, **15**:17. – (Type, in British Museum).
1913 *Aegocera tigrina* aberr. *albifascia* JORDAN in SEITZ, Großschmetterl. Erde, **15**:17. – (Type, in British Museum).
1913 *Aegocera tigrina* aberr. *flavifascia* JORDAN in SEITZ, Großschmetterl. Erde, **15**:17. – (Type, in British Museum).

Diagnosis: Head and thorax black; palpi with the first joint, the extremity of the second and the base of the third white; head and thorax with white dots; pectus with orange pilosity; tibiae and tarsi spotted with white; hind tibiae fringed with orange hair; abdomen black, with the three first urites orange, and the remaining urites with white edges; basal crest black. Fore wings black; an oblique yellow or yellowish-white subbasal line to vein 1; other markings yellow or yellowish-white: a rather broad oblique inner line; a short line at the extremity of cell; a broad, especially towards the tornus, outer line, oblique, followed on tornus by a small spot; a dark chocolate-red terminal band; between the yellow markings, silvery-blue lines and spots; cilia black with white spots in dorsal area. Hind wings orange-red, with a black terminal band, about 4 mm wide; cilia spotted with black and white. Length of fore wing 22–27 mm.

Distribution: From the Niger delta to Angola and south Zaire.

Polacanthopoda tigrina tigrina (DRUCE)

1913 *Aegocera tigrina tigrina* – JORDAN in SEITZ, Großschmetterl. Erde, **15**:17; pl. 4d.

Diagnosis: Yellow markings and cilia as above.

Distribution: From the Niger delta to Lower Congo River.

Polacanthopoda tigrina fabricata (KARSCH)

1895 *Mitrophrys fabricata* KARSCH, Entomol. Nachr.Bl., Wien, **21**:355; pl. 2, fig. 4. – Terra typica: Lower Guinea (Type, ♂, in Museum Berlin).
1912 *Rothia contigua* aberr. *fabricata* – STRAND, Lepidopt. Cat., **5**:19.
1913 *Aegocera tigrina fabricata* – JORDAN in SEITZ, Großschmetterl. Erde, **15**:17.

Diagnosis: Fore wings with the yellow markings reduced; cilia on both wings without white, or with white much reduced.

Distribution: Angola, Zaire: Uele, Tshuapa, Kasai, Kivu, Shaba. Specimens from Uganda differ in the shape of the inner line and may belong to a distinct subspecies.

Polacantopoda humphreyi HAMPSON

1911 *Polacanthopoda humphreyi* HAMPSON, Ann. Mag. nat. Hist., London, (8) 8:413. – Terra typica: Ilesha, South Nigeria (Type, ♂, in British Museum).
1913 *Aegocera humphreyi* – JORDAN in SEITZ, Großschmetterl. Erde, **15**:17; pl. 4c.
1920 *Polacanthopoda humfreyi* HAMPSON (ex errore pro *humphreyi*), Cat. Lepidopt. Phal. Brit. Mus., Suppl. **2**:563.

Diagnosis: Head black, white at sides; palpi white at base; the extremity of second joint and base of third white; antennae with the shaft white above; thorax deep chocolate-red with the dorsum black, the collar with white stripes; the thorax with white dorsal streak; the tegulae with white stripes and orange hair on upper edge; pectus orange with some white in front; fore legs black with orange streaks on femora and tibiae, the latter fringed with white on outer side; the mid and hind legs orange, the mid tibiae with black spots near base and extremity and a white ring at extremity; the hind tibiae with black spot near extremity, and a white ring at extremity; the tarsi black; abdomen orange with a black dorsal stripe and some silvery scales in the crests, a black segmental line on fourth segment, the three terminal urites black ringed with white; the underside with the four terminal urites black spotted with white. Fore wings deep chocolate-red; the costal area blackish to beyond middle, the terminal area redder; an oblique white striga from base of costa; a very oblique subbasal yellowish white line from costa to submedian fold with a silvery spot before it below costa and a line from cell to vein 1; an oblique white line sprinkled with black from below costa before middle to below middle of cell with a silvery mark below it above vein 1; some silvery below antemedial part of costa; an oblique bar across middle of cell; an oblique white bar across end of cell; a silvery discocellular bar with some silvery above it below costa; a yellowish spot below vein 2; a silvery striga at dorsum; an oblique postmedial series of white streaks sprinkled with black in the interspaces from costa to vein 3, followed by a white sprinkling to dorsum, with a silvery line before it, followed by an oblique white line from costa to vein 2, dentate between the streaks towards its lower extremity, then by a silvery line from costa to vein 4; a subterminal series of fine whitish streaks sprinkled with blackish on each side of the nervures. Hind wings deep orange, with a broad black terminal band, its inner edge incurved below vein 2; cilia chequered black and white. Length of fore wing 23 mm.

Distribution: Nigeria.

Polacanthopoda anthina (JORDAN)

1926 *Aegocera anthina* JORDAN, Novit. Zool., **33**:375–376. – Terra typica: Bauchi Plateau, North Nigeria (Type, ♀, in British Museum).

Diagnosis: Antennae and frons creamy; palpi pale yellow, spot on second and extremity of third joint black; thorax orange yellow, on each side a black spot on pronotum, and a lateral and dorsal stripe on meso- and metanotum; pectus and legs pale orange; tibiae and tarsi spotted with black; abdomen pale orange, with a black basal tuft. Fore wings dark brown; main markings buffy yellow; a patch in basal area, from below costa to dorsum, broader at middle, angulate distally below cell; in middle of patch below cell a black dot; proximally, a silvery double spot and two yellow spots at costal margin; a silvery line and a ferruginous red patch between cell and dorsum; discocellular mark buffy yellow, edged with black, with proximally and distally a silvery line; beyond that, a large more or less rounded buffy yellow costal patch, edged with black, edged distally with a silvery line continuing to dorsum and connected with the discocellular bar; further distal, a second silvery line, with space in between ferruginous red; terminal area pale greyish brown with silvery sheen, tinged proximally with ferruginous red; cilia buffy yellow. Hind wings ochraceous; a blackish brown discocellular mark and terminal band 5 mm wide; extremities of cilia buffy yellow. Length of fore wing 22–24 mm.

Distribution: Northern Nigeria.

Polacanthopoda naveli (LE CERF)

1922 *Aegocera naveli* LE CERF, Bull. Mus. Hist. nat., Paris, **28**:60. – Terra typica: Ile de Prince. (Type, sex undetermined, in Paris Museum).

Diagnosis: Head and abdomen missing. Body much damaged. Thorax above and tegulae chocolate brown, narrowly edged with white in front; pectus creamy white; legs brown with creamy white pilosity. Fore wings chocolate brown, paler in terminal area; from base to cell-extremity, a triangular creamy white patch, occupying the cell and the basal part of interspace II; a minute chocolate brown dot in middle of cell, followed by a larger rounded spot; a postmedial oblique creamy white band from below costa at two-thirds to vein 3, very faintly incurved distally, and rounded below; between the basal patch and the postmedial band, two pale transversal lines from below costa to end of cell; cilia concolorous. Hind wings chocolate brown with an orange patch from base, above, in, beyond and below cell, broadly excurved and slightly waved distally, rather diffuse towards the anal margin; a small brown cell-spot in it; cilia whitish with a fine brown median line. Length of fore wing 18 mm.

Distribution: Sao Thomé.

Genus **Aethodes** HAMPSON

(fig. 14)

1918 *Aethodes* HAMPSON, Novit. Zool., **25**:216. – Species typica: *Aethodes angustipennis* HAMPSON (by original designation).

Diagnosis: Proboscis fully developed. Antennae slightly dilated towards the extremity. Palpi very long, obliquely upturned, with the second joint reaching to about middle of frons and fringed with long hair in front produced to a point at the extremity; the third joint long and thickly scaled below. Frons with a small pointed conical process at middle. Tibiae fringed with short hair and without spines. Abdomen with hairy crests on the basal segments. Fore wings very narrow. Nervation: in fore wings, vein 2 given off from two-thirds of cell; 3 and 4 well separated; 4 and 5 approximated; discocellulars faintly incurved; 6 from the upper cell-angle; areole very narrow; 7 and the stalk of 8 + 9 from ite extremity; 10 from its upper angle. In hind wings, veins 2 from three-fifths of cell; 3, 4 and 6, 7 from the cell-angles; discocellulars incurved; 5 obsolescent, from the middle of the discocellulars; 8 anastomosing with cell at a point near the base.

Male genitalia: Uncus relatively small, narrow, pointed, faintly arched. Tubus analis broadening after base, but ending in a much narrower terminal portion; dorsal side well sclerified. Fold of vinculum short, triangular, bearing a broad bunch of stiff hair. Valva rather broad; costa bearing just beyond the middle a rounded

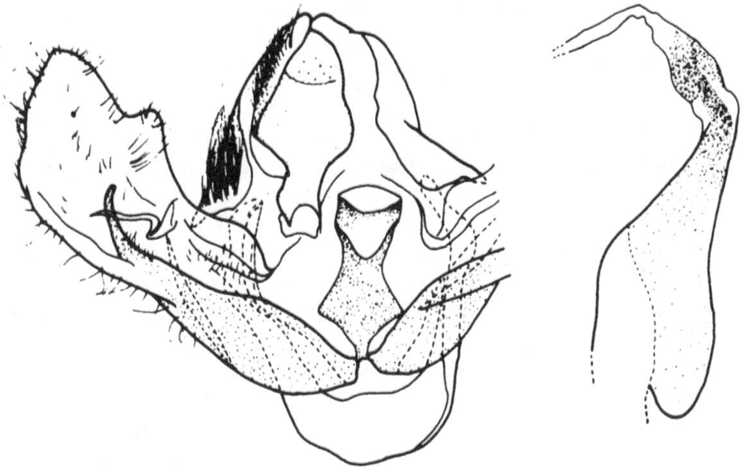

Fig. 14. *Aethodes angustipennis* HAMPSON. B. M. Noct. 8460. Typus

prominence; termen oblique, very faintly outcurved; fold of sacculus about half as broad as the valva proximally, becoming still narrower, reaching to the middle of valva, and bearing terminally a short, blunt, arched harpe. Aedeagus about as long as the valva, robust in basal two-fifths, then becoming much more slender, arched beyond two-thirds and bearing in the narrow area a proximal patch of larger, and a subterminal ring of minute cornuti; a terminal very slender process seems to be partly broken off in the type specimen; fultura inferior lozenge-shaped, with the distal end bearing two divergent prongs. Saccus very broadly rounded.

D i s t r i b u t i o n : West Africa.

Aethodes angustipennis HAMPSON

1918 *Aethodes angustipennis* HAMPSON, Novit. Zool., **25**:216. – T e r r a t y p i c a : Minna, North Nigeria (Type, ♂, in British Museum).

D i a g n o s i s : Head and thorax purplish-red mixed with white; antennae brown; frons white; palpi white with a band on the second joint, and the whole third joint, dark brown; pectus and hind legs yellowish; fore and mid legs banded brown and white; abdomen yellow with blackish-brown crests. Fore wings purplish-red, thickly sprinkled with white; a diffused inwardly-oblique white antemedial band edged distally with deeper red, and patches in and below cell; a deep red discocellular patch with a crimson line from it to dorsum; postmedial line deep red, rather diffused; a slightly dentate deep red subterminal shade with a small spot beyond it at the discal fold; apex slightly tinged with red. Hind wings pale orange yellow; termen narrowly pale red on apical half. Length of fore wing 17 mm.

D i s t r i b u t i o n : Nigeria; Gold Coast.

Genus **Aegoceropsis** KARSCH

(fig. 15)

1895 *Aegoceropsis* KARSCH, Entomol. Nachr.Bl., Wien, **21**:348. – *Aegoceropsis norma* KARSCH = *Aegocera affinis* DRUCE (by original designation).

D i a g n o s i s : Proboscis fully developed; palpi upturned, the second joint fringed with long hair in front, produced at the extremity to an acute angle; the third joint long, naked, somewhat porrect; frons with a truncate conical prominence, with a raised rim at the extremity; antennae more or less dilated towards the extremity; tibiae fringed with hair above; mid and/or hind tibiae with spines on the inner face; abdomen with a dorsal crest on the first urite; terminal urites in ♂ with lateral tufts of long hair, and protrusible tufts from base. Fore wings in ♂ with sometimes a trace of a groove below costa beyond middle, and the veins slightly curved downwards. Nervation: in fore wings, vein 2 given off at about three-quarters of cell; 3, 4, 5 approximated, equidistant; discocellulars incurved; 6 from the upper cell-angle; areole short; 7 and stalk of 8 + 9 from its extremity; 10 from its upper angle. In hind wings, vein 2 given off at three-fifths to three-quarters of cell; 3, 4 and 6, 7 from the cell-angles; discocellulars faintly incurved; 5 obsolescent, from middle of discocellulars; 8 anastomosing with cell for a short while near base.

Male genitalia: Uncus typically very small, but mostly moderate, narrowing towards the extremity. Tubus analis short and broad. Fold of vinculum moderately broad, fringed with rather short stiff hair. Valva well developed, banana-shaped, with the termen rounded; fold of sacculus rather broad proximally, narrowing to reach to middle of valva and ending in a rather robust, arched harpe, typically faintly dentate in middle of the incurved side. Aedeagus somewhat longer than valva, rather robust, produced distally in a narrow process, unarmed; fultura inferior nearly parallelogramm-shaped, with the sides faintly incurved. Saccus squarish with the angles rounded.

D i s t r i b u t i o n : Practically whole of the Ethiopian faunistic region.

Key to the species

1 Fore wings with a single broad whitish stripe from base to near termen 2
1′ Fore wings with spots in addition to the stripe 3
2 Whitish stripe continuous, with black spots round its middle . . . . *A. rectilinea*, p. 62
2′ Whitish stripe interrupted on discocellulars . . . . *A. brevivitta*, p. 62
3 Fore wings without pale stripe; dorsal area dark red . . . . *A. tricolora*, p. 157
3′ Fore wings with pale stripe below cell 4
4 Fore wings ferruginous red, without subapical pale oblique band . . . . . . *A. ferrugo*, p. 63
4′ Fore wings dark brown or ferruginous-red-brown, but with a subapical pale oblique band 5
5 Fore wings with the longitudinal stripe and the subapical band meeting above the anal vein and forming an angle . . . . *A. geometrica*, p. 62
5′ Fore wings with the longitudinal stripe and the subapical band not meeting 6
6 Fore wings with the subapical band very oblique, almost reaching termen at middle . . . *A. obliqua*, p. 61
6′ Fore wings with the subapical band less oblique, almost reaching termen above tornus . 7
7 Abdomen with a series of dorsal dark spots . . . . *A. fervida*, p. 59
7′ Abdomen without dark dorsal spots . . . . *A. affinis*, p. 58

Aegoceropsis affinis (DRUCE)

1883 *Aegocera affinis* DRUCE, Entomologist's mon. Mag., **20**:155. – T e r r a t y p i c a : East Africa (Type, ♂, in British Museum).

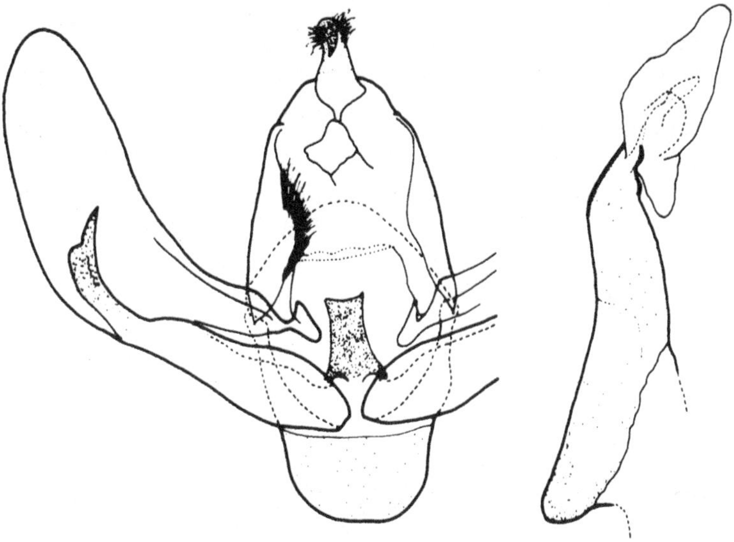

Fig. 15. *Aegoceropsis affinis* (DRUCE). B. M. Noct. 8292. Typus

1895 *Aegoceropsis norma* KARSCH, Entomol. Nachr.Bl., Wien, **1895**:352; pl. 2, fig. 3. – Terra typica: Manboia, East Central Africa (Type, ♂, in British Museum).
1913 *Aegocera fervida affinis* – JORDAN in SEITZ, Großschmetterl. Erde, **15**:18; pl. 3g.

Diagnosis: Palpi yellow; the sides of second joint and tip of third black; head cream-coloured; the sides of frons and antennae black; thorax above rufous with lateral and dorsal white stripes; pectus and legs orange yellow; fore and mid tibiae and tarsi ringed with black; abdomen orange yellow, unspotted. Fore wings ferruginous-red-brown; costa whitish to the post-medial band, with a streak of black scales on it and a leaden subbasal dot; a whitish fascia from base of costa to well beyond middle in the submedian interspace; an oblique whitish band from two-thirds of costa to above vein 2 near termen; a purplish-leaden spot in middle of cell and a discocellular bar; cilia whitish. Hind wings orange yellow, with a black discocellular lunule; a terminal band, red-brown in ♂ and blackish in ♀, with sinuous inner edge, about 3 mm wide at apex, and narrowing to tornus. Length of fore wing 20 mm.

Distribution: Tanzania, Zambia, Malawi.

Aegoceropsis fervida (WALKER)

1854 *Aegocera fervida* WALKER, List Lepidopt. Het. Brit. Mus., **1**:57. – Terra typica: Natal (Type, ♂, in British Museum).
1901 *Aegocera (Mitrophrys) fervida* – HAMPSON, Cat. Lepidopt. Phal. Brit. Mus., **3**:601.

Diagnosis: Head and thorax blackish brown; palpi fringed with orange hair; extremity of the second and base of the third joint white; frons, vertex, collar, and thorax above with a broad whitish stripe; pectus and legs orange; fore tibiae and tarsi ringed with black; abdomen orange, with dorsal series of black spots. Fore wings deep chocolate-red in ♂, blackish-brown in ♀, sparsely sprinkled with ochreous, especially on the costal area; a cream-coloured fascia from base of costa to well beyond middle in the submedian interspace; an oblique fascia from costa beyond middle to above vein 2 near termen; a leaden spot below costa near base; an antemedial bar below the fascia, and one beyond the same; a bar in middle of cell, another on the disco-cellulars, all leaden-coloured; cilia cream-coloured. Hind wings orange; a discocellular spot and a terminal band, rusty-red in ♂ and blackish in ♀; cilia orange. Length of fore wing 18–26 mm.

Distribution: South and East Africa to Angola, Zaire and Kenya.

Aegoceropsis fervida fervida (WALKER)

1860 *Aegocera triphaenoides* WALLENGREN, Wien. entomol. M.Schr., **4**:38. – Terra typica: Caffraria or. (Type in Museum Stockholm).
1865 *Aegocera fimbria* MÖSCHLER, Stettin. entomol. Ztg., **33**:354. – Terra typica: Silhet (ex errore). – Terra typica restricta (hoc loco): Cape Province.
1913 *Aegocera fervida* ssp. *fervida* – JORDAN in SEITZ, Großschmetterl. Erde, **15**:18; pl. 3g.

Diagnosis: Hind wings with the terminal band excised at middle and angled inwards on vein 2.

Distribution: Republic of South Africa (Cape Province, Transvaal, Natal).

Aegoceropsis fervida fuscicosta (JORDAN)

1913 *Aegocera fervida* ssp. *fuscicosta* JORDAN in SEITZ, Großschmetterl. Erde, **15**:18. – Terra typica: Lilongwe, Nyasaland (Type, ♂, in British Museum).

Diagnosis: Abdomen with black spots obsolescent in ♂; sides of basal crest black in ♂, fore wings with the costal streak white, densely sprinkled with black.

Distribution: Malawi to Uganda (Kigezi).

Aegoceropsis fervida media (JORDAN)

1913 *Aegocera fervida* ssp. *media* JORDAN in SEITZ, Großschmetterl. Erde, **15**:18. – Terra typica: Angola, Cauhoca (Type, ♂, in British Museum).

Diagnosis: Abdomen with the black spots large, distal spots tending to form bands; hind wings with the terminal band narrower; cilia at apex generally darkened.

Distribution: Angola; Zaire (Kinshasa, Kasai, Equateur); Burundi; Congo (Brazzaville); Cameroon.

Aegoceropsis obliqua (MABILLE)

1893 *Aegocera obliqua* MABILLE, Ann. Soc. entomol. Belg., Bruxelles, **37**:56. – Terra typica: Loko, West Africa (Type, ♂, in Berlin Museum).
1901 *Aegocera (Mitrophrys) obliqua* – HAMPSON, Cat. Lepidopt. Phal. Brit. Mus., **3**:602.

Diagnosis: Head and thorax deep chocolate-red; palpi fringed with orange hair; the third joint white, black at the extremity; head and thorax with a medial white stripe; shaft of the antennae white dorsally except towards base; pectus and legs orange; fore and mid tibiae and tarsi banded black and white; abdomen orange, with a series of dorsal black spots. Fore wings deep chocolate red; a yellowish white stripe from base of costa to near termen in the submedian interspace, interrupted at middle, and with a small rounded spot separated off at the extremity; an oblique band from below costa beyond middle to near termen above vein 3; a few bluish silvery scales below costa before middle, and others near termen; cilia orange. Hind wings orange with a deep red or black discocellular spot and broad terminal band, expanding on apical and submedian interspace; cilia orange, brownish towards the apex. Length of fore wing 19–22 mm.

Distribution: West and Central Africa to Uganda.

Aegoceropsis obliqua obliqua (MABILLE)

1913 *Aegocera obliqua* ssp. *obliqua* – JORDAN in SEITZ, Großschmetterl. Erde, **15**:18; pl. 3g.

Diagnosis: Postdiscal band on fore wings rather narrow; hind wings with markings red in ♂, black in ♀.

Distribution: Sierra Leone to Nigeria.

Aegeceropsis obliqua isogenis (JORDAN)

1913 *Aegocera obliqua* ssp. *isogenis* JORDAN in SEITZ, Großschmetterl. Erde **15**:18; pl. 3g ♂ (as *obliqua* ♀). – Terra typica: Bopote, Congo (Type, ♂, in British Museum).

Diagnosis: Fore wings with the postdiscal band wider; hind wings with the markings red in both sexes.

Distribution: Zaire (Uele, Ituri, Equateur, Tshuapa), Cameroon.

Aegoceropsis obliqua remutata (JORDAN)

1913 *Aegocera obliqua* ssp. *remutata* JORDAN in SEITZ, Großschmetterl. Erde, **15**:18. – Terra typica: Campala, Uganda (Type, ♂, in British Museum).

Diagnosis: Fore wings with the postmedial band wider; hind wings with the markings red in ♂, black in ♀.

Distribution: Uganda, East Zaire: Maniema, Lake Tanganyika, Kivu; Burundi; Rwanda.

Aegoceropsis geometrica (HAMPSON)

1910 *Aegocera geometrica* HAMPSON, Proc. zool. Soc. London, **1910**:398; pl. 36, fig. 15. – Terra typica: Kambove District, Congo = Zaire (Type, ♀, in British Museum).

Diagnosis: Head white, black laterally; palpi chocolate red, fringed with yellow and brown hair; extremity of the second joint white; the third joint white at base and black at extremity; antennae black with some white above; thorax chocolate red with white patches on shoulders and metathorax; pectus and legs orange; fore and mid tibiae with some black above; terminal tarsal joint black above; abdomen orange with a dorsal black patch at base. Fore wings dark chocolate red; costa black sprinkled with white; costal edge white except towards base; an oblique white streak from below base of the costa to vein 1 beyond middle where it is met by a narrow white band from costa; cilia white. Hind wings orange with a broad black terminal band; cilia yellow at extremities. 15–17 mm.

Distribution: Zaire (Sankuru, Lulua, Shaba).

Aegoceropsis rectilinea (BOISDUVAL)

1836 *Aegocera rectilinea* BOISDUVAL, Spéc. gén. Lépidopt., **1**: pl. 14, fig. 5. – Terra typica: West Africa: Galam (Type not designated).
1923 *Aegocera rectilinea* aberr. *dealbata* HULSTAERT, Rev. Zool. Bot. Afr., Bruxelles, 000:000. – Terra typica: Nyangwe, "Congo" (Type, ♀, in Tervuren Meseum).

Diagnosis: Head and thorax white; antennae and sides of the second and extremity of the third palpal joint black; collar striped with rufous and black; tegulae edged with blue black and orange; thorax above with sublateral blue black and orange stripes; tibiae and tarsi orange, banded with black and white; abdomen orange with a dorsal series of black spots; underside whitish. Fore wings dark brown, very thickly sprinkled with ochreous; a metallic blue spot below base of costa; a creamy white fascia edged with rusty-red from base along the media and below vein 6, ending in a point just before termen, its upper edge indented by silvery-blue cell- and discocellular spots, and its lower edge by a spot before the triangular expansion at the cell-extremity; terminal line orange-red; cilia blackish tipped with whitish. Hind wings orange; discocellular spot and terminal band rusty-red in ♂, black in ♀; terminal line orange; cilia dark brown, tipped with whitish. Length of fore wing 14,5–19 mm.

Distribution: Most of tropical Africa, to Oubangui-Chari, Sudan, Egypt. Ethiopia, the whole of Zaire, Burundi, also Kenya and Tanzania.

Aegoceropsis brevivitta (HAMPSON)

1901 *Aegocera (Mitrophrys) brevivitta* HAMPSON, Cat. Lepidopt. Phal. Brit. Mus., **3**:600; pl. 52, fig. 10. – Terra typica: Athi-ya-Mawe, British East Africa = Kenya (Type, ♀, in British Museum).
1913 *Aegocera brevivitta* – JORDAN in SEITZ, Großschmetterl. Erde, **15**:17; pl. 4d.
1913 *Aegocera irangiana* WICHGRAF, Internation. entomol. Z., Guben, **7**:14. – Terra typica: Kondoa Irangi (Type, ♀, in ?Museum Berlin).
1937 *Aegocera brevivitta* forma *triangularis* NIEPELT, Festschr. E. STRAND:559; fig. 2. – Terra typica: Somalia, Afgoi (Type, ♀, in Museum Berlin).

D i a g n o s i s : Head and thorax white; sides of the second palpal joint rufous; tuft at the extremity of the third, and antennae black; collar striped rufous and black; edges of tegulae and subdorsal stripes on thorax black and orange; tibiae and tarsi orange, banded with black and white; abdomen orange with a dorsal series of black spots and sublateral points; underside white. Fore wings dark brown, very thickly sprinkled with ochreous; a white fascia from base along the media, expanding into a hammerhead at cell-extremity, and edged except above with rusty-red; its upper and lower edge slightly indented before the extremity by red and silvery-blue spots and divided by a red and silvery-blue reniform stigma from a red-edged wedgeshaped white spot beyond cell below vein 6, ending some distance from termen; terminal line red; cilia blackish, tipped with whitish. Hind wings deep ferruginous with yellowish base in ♂, orange or yellow in ♀, the latter having a small blackish discocellular lunule and a terminal blackish band, diffuse on inner area towards the base, a broad costal fascia to before middle; terminal line orange; cilia blackish, tipped with whitish in ♂, with yellowish in ♀. Length of fore wing 14–16 mm.

D i s t r i b u t i o n : Kenya, Tanzania, Angola; Air.

A e g o c e r o p s i s b r e v i v i t t a b r e v i v i t t a (HAMPSON)

D i a g n o s i s : Hind wings orange, as above.

D i s t r i b u t i o n : As above except Air.

A e g o c e r o p s i s b r e v i v i t t a r e c t i l i n e o i d e s (ROTHSCHILD)

1921 *Aegocera rectilineoides* ROTHSCHILD, Novit. Zool., **28**:158. – T e r r a t y p i c a : Damergou, Air (Type, ♂, in British Museum).
1934 *Aegocera brevivitta* forma *rectilineoides* – GAEDE in SEITZ, Großschmetterl. Erde, **15**:27.

D i a g n o s i s : Hind wings mostly yellow, with the discocellular mark and the outer area rusty-red.

D i s t r i b u t i o n : Air; Nigeria (fide GAEDE in SEITZ 1934).

Aegoceropsis ferrugo (JORDAN)

1926 *Aegocera ferrugo* JORDAN, Novit. Zool., **33**:375. – T e r r a t y p i c a : Hora Daka, Ethiopia (Type, ♂, in British Museum).

D i a g n o s i s : Head and thorax cream-coloured; first and second palpal joints chestnut with some blackish and creamy on outside; fringes orange, mixed on outside with black; third joint black; stripe from eye to pronotum and edges of tegulae chestnut red; pectus orange buff; fore and mid tibiae mixed grey with orange buff pilosity; hind tibiae greyish black with orange buff basal pilosity; tarsi black spotted with white; abdomen orange buff; bases of tergites III to VII black; underside ochreous. Fore wings ferruginous-red; costal and dorsal edges sprinkled with creamy; a diffuse stripe from base along the submedian fold to near termen and a diffuse terminal line cream-coloured, sprinkled with black; near base of cell an obsolescent creamy spot mixed with black; before cell-extremity an oblong spot, behind cell-extremity a lunule, both cream-coloured edged with black; a partly scaleless stripe along costa beyond middle; cilia yellowish. Hind wings slightly more yellowish; spurs of a tawny discocellular spot; long basal and anal hair and cilia yellow. Length of fore wing 16,5 mm.

Distribution: Ethiopia.

Genus **Mitrophrys** KARSCH

(fig. 16)

1895 *Mitrophrys* KARSCH, Entomol. Nach.Bl., Wien, **21**:349. – Species typica: *Noctua menete* CRAMER (by original designation).

Diagnosis: Proboscis fully developed; antennae faintly dilated towards the extremity, nearly setiform; palpi upturned, with the second joint bearing a fringe of long hair, especially at tip, and the third joint long, smooth; frons with a truncate conical prominence with a raised rim at the extremity, not narrowed towards the hind margin in male, clothed with scales and appressed hair; abdomen with a basal dorsal crest; in male, with protrusible tufts of hair from base, and with lateral tufts of long hair on terminal urites. Fore wings of male without subcostal groove; hind wings without fringe of long hair on anal area. Venation: in fore wings, vein 2 from slightly beyond two-thirds of cell; 3, 4, 5 well separated, equidistant; discocellulars incurved; 6 from lower cell-angle; areole small; 7 and the short stalk of 8 + 9 from the tip of areole; 10 from upper margin of areole, near the extremity. In hind wings, vein 2 from three-quarters of cell; 3, 4 and 6, 7 from the cell-angles; discocellulars faintly incurved; 5 obsolescent, from middle of discocellulars.

Male genitalia: Uncus very small, narrow, angled before middle, tapering to an acute point; pilosity scarce. Vinculum rather narrow; fold narrow, bearing a bunch of stiff hair. Valva banana-shaped, moderately broad, with the apex rounded; fold of sacculus moderately broad, reaching to two-thirds of sacculus and ending in a rather small, arched harpe. Aedeagus slightly longer than valva, boomerang-shaped, moderately robust and somewhat slenderer in terminal half; the latter bearing an elongate patch of minute cornuti; fultura inferior with the proximal margin produced at middle, incurved sublaterally, and with the angles produced and acute; sides incurved; terminal margin deeply excised at middle. Saccus semi-circular.

Distribution: Africa.

Key to the species

1 Fore wings black or blackish brown in both sexes 2
1′ Fore wings rusty-red in male, dark red brown or blackish brown in female 3
2 Fore wings with the postmedial band oblique; hind wings with the terminal band red brown . . . . *M. barnsi*, p. 69
2′ Fore wings with the postmedial band vertical; hind wings with the terminal band blackish brown in both sexes . . . . *M. gynandra*, p. 68
3 Fore wings with a whitish subterminal line . . . . *M. ansorgei*, p. 69
3′ Fore wings without whitish subterminal line 4

4 Fore wings with the postmedial band vertical . . . . *M. latreillii*, p. 68
4′ Fore wings with the postmedial band oblique 5
5 Antennae white, slightly marked with black . . . . *M. menete*, p. 65
5′ Antennae black; some white scales in male . . . . *M. magna*, p. 66

Mitrophrys menete (CRAMER)

1775 *Noctua menete* CRAMER, Pap. Exot., **1**:pl. 70 D. – Terra typica: West Africa (Type not designated).
1787 *Noctua formosa* FABRICIUS, Mant. Ins., **2**:150. – Terra typica: Africa (Type not designated).
1854 *Aegocera magna* WALKER (partim), List Lepidopt. Het. Brit. Mus., **1**:56. – Terra typica: Sierra Leone (Types, ♂, ♀, in British Museum).
1874 *Aegocera menete* – BOISDUVAL, Rev. Mag. Zool., (3) 2:52.
1895 *Mitrophrys menete* – KARSCH, Entomol. Nachr.Bl., Wien, **21**:349.

Diagnosis: Head and thorax rufous in male, blackish brown in female; antennae white, frequently spotted with black laterally; 2nd joint of palpi, lateral stripes on frons, vertex, sides and middle of collar, upper edge of tegulae and stripe on middle of thorax, yellowish in male, nearly white in female; pectus and legs orange; fore and mid tibiae and tarsi with black spots; abdomen orange, with a

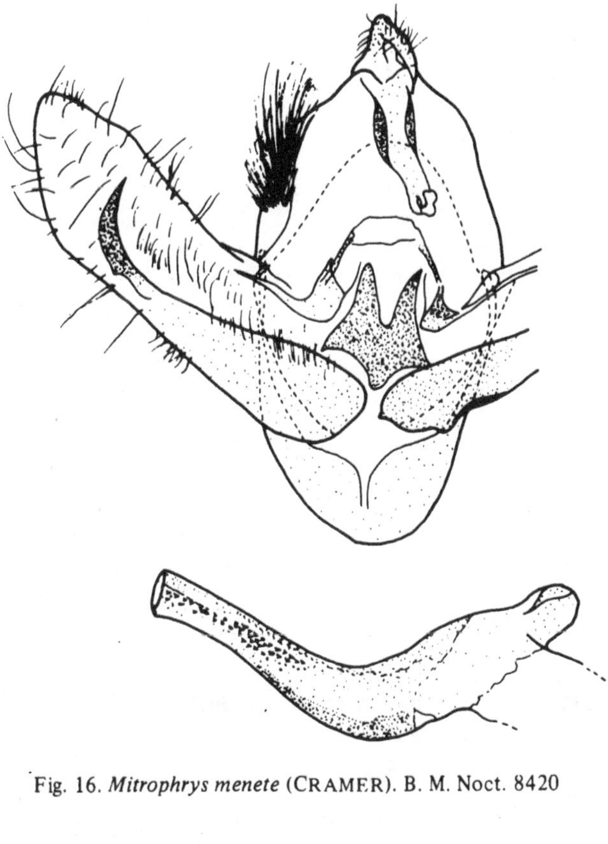

Fig. 16. *Mitrophrys menete* (CRAMER). B. M. Noct. 8420

dorsal series of black spots. Fore wings rusty red in male, rufous brown in female; markings yellowish in male, nearly white in female, viz.: a spot below base of costa, followed by two elongate spots, and then by two rounded spots in submedian interspace; a spot at cell-end; an oblique band of two conjoined spots from costa at two-thirds to interval II; some leaden scales below costa near base; antemedial spots in cell and in interval I; a discocellular bar; cilia yellow. Hind wings orange; markings dark rusty brown in male, blackish brown in female, viz. a large discocellular bar, often much reduced in female; a broad terminal band, excurved in interval I; cilia yellow. Length of fore wing 20–26 mm.

Distribution: Senegambia to Eritrea; Central and East Africa.

Mitrophrys menete menete (CRAMER)

1913 *Mitrophrys menete menete* – JORDAN in SEITZ, Großschmetterl. Erde, 15:19; pl. 3 h.

Diagnosis: As above.

Distribution: Senegambia to Nigeria, Sierra Leone, Gold Coast, Old Calabar, Kavirondo, Uganda, Ruwenzori, Sudan, Tanzania, Kenya.

Mitrophrys menete rubida (FELDER)

1874 *Aegocera rubida* FELDER, Reise Novara, 1:pl. 107, fig. 14. – Terra typica: West Africa (Type, ♂, in British Museum).
1913 *Mitrophrys rubida* – JORDAN in SEITZ, Großschmetterl. Erde, 15:19; pl. 3h.

Diagnosis: Fore wings with the costal area yellowish.

Distribution: Ethiopia and Eritrea.

Mitrophrys magna (WALKER)

1854 *Aegocera magna* WALKER, List Lepidopt. Het. Brit. Mus., 1:56. – Terra typica: Sierra Leone (Types, ♂, ♀, in British Museum).
1913 *Mitrophrys magna* – JORDAN in SEITZ, Großschmetterl. Erde, 15:19; pl. 4b and c.

Diagnosis: Head and thorax rufous in male, blackish brown in female; antennae black, with a few white scales, sometimes with a brown stripe; extremity of the second palpal joint, base of the third, vertex, spots on collar, edge of tegulae, stripe on thorax, whitish; pectus and legs orange; tarsi spotted with black; abdomen orange with a dorsal series of black spots; anal hair long; underside white. Fore wings rusty red in male, brown in female; a discal series of pale yellowish spots reaching more distally in the submedian interspace; a discocellular spot; an oblique band of two conjoined spots from costa at two thirds to interval II; some leaden scales below costa near base; antemedial spots in cell and above vein 1; a small discocellular mark; terminal band with a tooth on the lower media, tapering to a point at anal angle. Length of fore wing 26 mm.

Distribution: West and Central Africa to Kenya and Tanzania.

Mitrophrys magna magna (WALKER)

Diagnosis: As above.

Distribution: Senegambia, Sierra Leone, Gold Coast, Nigeria.

Mitrophrys magna mesia (JORDAN)

1913 *Mitrophrys mesia* JORDAN in SEITZ, Großschmetterl. Erde, **15**:19; pl. 4c. – Terra typica: Unyoro (Type, 000, in British Museum).

Diagnosis: Frons whitish with a mesial brown streak; streak in submedian interspace excised once or twice dorsally near its extremity; hind wings orange yellow, sometimes tinged with rusty red; terminal band broad, rufous brown, excurved at middle.

Distribution: Unyoro, Zaire; Uganda.

Mitrophrys magna obliquisigna (HAMPSON)

1901 *Aegocera obliquisigna* HAMPSON, Cat. Lepidopt. Phal. Brit. Mus., **3**:603, pl. 54, fig. 13. – Terra typica: British East Africa (Type, ♂, in British Museum).
1909 *Aegocera kibwezensis* STRAND, Arch. Naturgesch., Berlin, **25** 1 (3):367. – Terra typica: Kibwezi, East Africa (Types, ♂♀, in Berlin Museum).
1909 *Aegocera kibwezensis* aberr. *aequipartita* STRAND, Arch. Naturgesch., Berlin **25** 1(3):367. – Terra typica: Kibwezi (Type, ♂, in Museum Berlin).
1913 *Mitrophrys obliquisigna* – JORDAN in SEITZ, Großschmetterl. Erde, **15**:20; pl. 3h.

Diagnosis: Near *M. magna mesia*, but smaller; frons without the mesial brown streak; hind wings more or less tinged with reddish brown. Length of fore wing 16 mm.

Distribution: Tanzania, Kenya, South Sudan.

Mitrophrys magna abyssinica (HAMPSON)

1901 *Aegocera abyssinica* HAMPSON, Cat. Lepidopt. Phal. Brit. Mus., **3**:604, pl. 54, fig. 14. – Terra typica: Atbara, Ethiopia (Types, ♂, ♀, in British Museum).
1913 *Mitrophrys abyssinica* – JORDAN in SEITZ, Großschmetterl. Erde, **15**:20.

Diagnosis: As *M. magna obliquisigna*, but abdomen with dorsal and lateral black bands; hind wings yellow, with the anal area rufous. Length of fore wing 16–19 mm.

Distribution: Ethiopia.

Mitrophrys magna kenyamagaribae STONEHAM

1963 *Mitrophrys kenya-magaribae* STONEHAM, Bull. Stoneham. Mus., Nairobi, **80**:pl. 2. – Terra typica: Istsare, Kenya (Type, ♂♀, in Stoneham Museum).

D i a g n o s i s : Fore wings with the postmedial band less oblique, but broad and long; hind wings with terminal band rusty red in male, black in female.

D i s t r i b u t i o n : West Kenya.

Mitrophrys latreillii (HERRICH-SCHÄFFER)

1853 *Aegocera latreillii* HERRICH-SCHÄFFER, Außereurop. Schmetterl.; fig. 19. – T e r r a t y p i c a : Casamanca, Senegambia (Type, ♀, in British Museum).
1913 *Mitrophrys latreillii* – JORDAN in SEITZ, Großschmetterl. Erde, **15**:20; pl. 3g and h.

D i a g n o s i s : Head, thorax and fore wings bright rufous in male, blackish brown in female; antennae black; palpi fringed with orange hair; extremity of the second joint, base of the third, sides of frons, lateral stripes on vertex, sides and middle of collar, streaks on tegulae and medial stripe on thorax, creamy white in male, white in female; pectus and legs orange; tibiae and tarsi banded with black; abdomen orange in male, paler, whitish in female, with dorsal black spots at base and bands towards extremity. Fore wings with a yellowish spot below base of costa, followed by an elongate spot, then a square, then a rounded, finally a triangular spot in submedian interspace; a round spot at cell-tip; postmedial band slightly oblique, from below costa to vein 3, constricted at middle; some purplish-silvery markings, viz., a point below base of costa; antemedial points in cell and in interval I; a discocellular bar; a postmedial line from below costa to vein 4, then incurved; a terminal series of conjoined spots; cilia yellow. Hind wings orange; markings rufous in male, blackish brown in female, viz., a discocellular spot, and a broad, strongly incurved at vein 2 terminal band; cilia orange. Length of fore wing 22 mm.

D i s t r i b u t i o n : Nyasaland, Mashona, Matabeleland, South Sudan.

Mitrophrys gynandra JORDAN

1913 *Mitrophrys gynandra* JORDAN in SEITZ, Großschmetterl. Erde, **15**:20; pl. 4b. – T e r r a t y p i c a : Bihé, Angola (Type, ♂♀, in British Museum).
1920 *Aegocera gynandra* – HAMPSON, Cat. Lepidopt. Phal. Brit. Mus., Suppl. **2**:575; pl. 71, fig. 19.

D i a g n o s i s : Head and thorax deep chocolate brown; frons creamy white, with a black mesial stripe; thorax dorsally with a creamy white stripe; antennae black with the basal joint creamy white; pectus and legs orange; tibiae banded with black; tarsi ringed with white; abdomen orange, with a dorsal series of black spots towards the base, and bands on the penultimate segment. Fore wings dark chocolate brown; a creamy white elongate spot near base of costa, its upper margin indented near base and angled in cell, followed by an elongate spot, then by a bar-shaped, finally by a wedge-shaped spot, all in the submedian fold, nearly reaching the termen; a round spot in cell-end; a broad oblique postmedial band from costa to interval II, narrower on costa, ending on the proximal side in an inward angle; some silvery markings, viz., a patch of scales below costa at base; a bar at cell-middle; a bar below the extremity of the first spot of the fascia; a curved line on discocellulars; a few scales beyond the postmedial band; a band beyond the third spot of the fascia; and some sprinkling on the termen; cilia creamy white. Hind wings orange; markings chocolate brown in both sexes; a discocellular bar; and a broad terminal band, excurved prosimally at middle, and incurved at vein 2; cilia pale yellow. Length of fore wing 20–23 mm.

Distribution: Central and South Zaire; Zambia.

Mitrophrys barnsi JOICEY & TALBOT

1921 *Mitrophrys barnsi* JOICEY & TALBOT, Bull. Hill Mus. Witley, **1**:161. – Terra typica: Lufira River, near Likasi, Congo Free State = Zaire (Type, ♂, in British Museum).
1955 *Aegocera jordani* KIRIAKOFF, Bull. Inst. r. Sci. nat. Belg., Bruxelles, **31** (69):3–5; fig. 2. – Terra typica: Lubumbashi, Congo = Zaire (Type, ♂, in Brussels Museum).

Diagnosis: Head and thorax black; antennae black with the base pale buff; palpi fringed with orange hair below; frons pale buff with a black mesial stripe; collar, tegulae and thorax above spotted with buff; pectus orange; legs blackish with orange pilosity; abdomen orange, with black segmental lines, broader at middle dorsally. Fore wings black; markings buff, viz., an oval subcostal spot at base; a narrow spot at cell-base, with a parallel band below it; a bar in cell-middle; a large square discocellular spot; a large spot in submedian interspace, followed by a narrow lunulate spot; a broad oblique subapical band of two conjoined spots; a few pale costal scales near apex; most of the buff markings more or less broadly edged with indian red; some purplish silvery markings on red ground, viz., a broad lunule in interval I near the termen; and a subterminal series of spots; cilia buff. Hind wings orange; a broad black discocellular spot; terminal band reddish brown, incurved in interval iv and reaching somewhat beyond the anal angle; terminal line black; cilia blackish, tipped with orange. Length of fore wing 21–23 mm.

Distribution: Zaire (Shaba).

Mitrophrys ansorgei (ROTHSCHILD)

1897 *Aegocera ansorgei* ROTHSCHILD, Novit. Zool., **4**:311; pl. 7, fig. 9. – Terra typica: Uganda, Mnani (Type, ♂, in British Museum).
1934 *Mitrophrys ansorgei* – GAEDE in SEITZ, Großschmetterl. Erde, **15**:27; pl. 3h.

Diagnosis: Head and thorax fox-red in male, reddish brown in female; palpi with yellow pilosity; 3rd joint whitish at base; sides of frons and vertex striped with white; collar marked with white; tegulae and thorax above striped with white, with tegulae fringed with orange above; pectus and legs orange; tibiae and tarsi banded with whitish and black; abdomen orange with fine black segmental lines; underside whitish with sublateral black spots. Fore wings fox-red in male, reddish brown in female; costal and dorsal areas thickly sprinkled with creamy white; markings creamy, viz., a spot below base of costa, followed by two streaks in submedian interspace to well beyond middle, with a short curved line beyond its extremity, and some sprinkling above it; a small antemedial cell-spot; a squarish spot in cell-end; an oblique sinuous postmedial band from below costa to interval II; a subterminal line, incurved below vein 3; some leaden silvery marks, viz., a streak above base of vein 1; a bar across middle of cell to submedian fascia; a discocellular bar with a creamy striga before it; a terminal series of small spots; cilia yellow. Hind wings orange, with markings rufous in male, black in female, viz., a discocellular bar, and a terminal band excised at middle and angled inwards on vein 2. Length of fore wing 19–21 mm.

Distribution: Uganda; Kenya.

Genus **Agoma gen. nov.**

(fig. 17)

Species typica: *Aegocera trimeni* FELDER (designatione hoc loco).

Diagnosis: Proboscis fully developed; antennae of ♂ reaching to about two-thirds of costa, hardly dilated terminally; antennae of ♀ filiform; palpi porrect, one-and-half times as long as the diameter of eye, with the second joint with a long

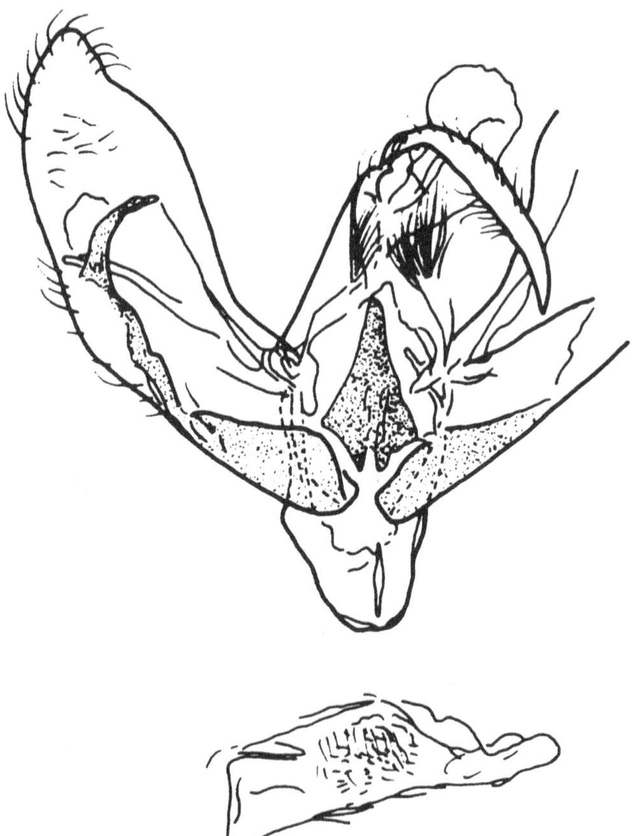

Fig. 17. *Agoma trimeni* (FELDER). T/72 Agar. 26

hanging hair-tuft, the third as long as the second, slender, clubbed at the extremity; frons with a rounded projection, with raised rim at the extremity; pilosity of legs moderate; hind tibiae in ♂ with a long hair-brush; hind tarsi very finely spined; abdomen with dorsal crests on the three proximal urites. Nervation: in fore wings, vein 2 approximated to the cell-angle; 3, 4, 5 approximated, with the distance 4–5 slightly shorter than the distance 3–4; discocellulars incurved; 6 from the lower cell-angle; areole short and narrow; 7 and stalk of 8 + 9 from its extremity; 10 from its upper margin. In hind wings, vein 2 given off at near the cell-angle; 3, 4 and 6, 7 from the cell-angles; discocellular very obsolete, evenly incurved; 5 obsolescent, from middle of the discocellulars; 8 fused to cell at base.

Male genitalia: Uncus reaching the base of valva when bent, arched near base, then very faintly bent, with a small terminal hook. Tubus analis slightly shorter than uncus, broad and expanding distally, with ventral sclerification. Vinculum rather narrow, fringed with stiff hair. Valva banana-shaped but rather broad; costa faintly outcurved in middle portion; saccus more distinctly outcurved; apex angulate; fold of sacculus rather broad basally, but very short, not reaching one-third of valva; harpe short, arched, placed beyond the middle of valva. Aedeagus slightly shorter than costa, very robust, spoon-shaped proximally, straight; vesica with a large patch of minute cornuti; fultura inferior nearly triangular with the apex rounded and with the proximal margin produced and notched at middle.

Distribution: Most of the Ethiopian faunistic region.

Agoma trimeni (FELDER)

1874 *Aegocera trimeni* FELDER, Reise Novara: pl. 107, fig. 15. – Terra typica: Natal (Type, ♂, in British Museum).
1901 *Tuerta trimeni* – HAMPSON, Cat. Lepidopt. Phal. Brit. Mus., 3:624.

Diagnosis: Head black; second palpal joint white at base and extremity; sides of frons and vertex striped with white; thorax blackish-brown; collar white in middle and on sides; tegulae and thorax above striped with white, the former fringed with orange above; pectus and legs orange; fore and mid tibiae and the tarsi banded with black; abdomen white, sometimes orange laterally, with a dorsal black stripe and lateral bands; underside white or black unspotted or spotted with white. Fore wings deep chocolate-brown, thickly sprinkled with pale olive-green; a wedge-shaped white fascia from base of costa to beyond middle in the submedian interspace, ending there in a point, its upper edge in cell excised; an elliptical or triangular band beyond cell from below costa to vein 3 on an olive-green patch; an oblique silvery striga before middle from the fascia to vein 1, with a red line on an olive-green band beyond it; markings in cell and on the discocellulars olive-green with a silvery bar in middle; a postmedial silvery line; a fine dark terminal line; cilia pale brown. Hind wings white, or white tinged with orange, or yellow, or orange to reddish-brown, with a black terminal band broad at costa and narrowing to the anal angle; cilia pale brown, tipped with white. Length of fore wing 17–24 mm.

Distribution: Most of the Ethiopian faunistic region.

Agoma trimeni trimeni (FELDER)

1883 *Aegocera tricolor* DRUCE, Entomologist's mon. Mag., **20**:155. – Terra typica: Delagoa Bay (Type, ♂, in British Museum).
1893 *Aegocera elegantula* MABILLE, Ann. Soc. entomol. Belg., Bruxelles, **37**:56. – Terra typica: Verulam, Natal (Type, ♀, in Berlin Museum).
1910 *Tuerta ovifera* HAMPSON, Proc. zool. Soc. London, **1910**:400; pl. 35, fig. 25. – Terra typica: Eastern Luangwa District, Northeast Rhodesia = Zambia (Type, ♀, in British Museum).
1913 *Aegocera trimeni* ssp. *trimeni* – JORDAN in SEITZ, Großschmetterl. Erde, **15**:18; pl. 2f.
1913 *Aegocera trimeni* ssp. *trimeni* forma *semialba* JORDAN in SEITZ, Großschmetterl. Erde, **15**:19.
1913 *Aegocera trimeni* ssp. *trimeni* aberr. *nigriventris* JORDAN in SEITZ, Großschmetterl. Erde, **15**:19.
1913 *Aegocera trimeni* ssp. *trimeni* aberr. *albiventris* JORDAN in SEITZ, Großschmetterl. Erde, **15**:19.

Diagnosis: Fore wings with a yellowish discal patch on the underside; markings above white or yellowish; hind wings white or orange.

Distribution: West, South and East Africa: Gold Coast, Lagos, Republic of South Africa (Cape Province, Transvaal, Natal), Zambia, Tanzania, Kenya, Moçambique, Ethiopia.

Agoma trimeni halans (KARSCH)

1895 *Mitrophrys halans* KARSCH, Entomol. Nachr.Bl., Wien, **21**:354; pl. 2, fig. 7. – Terra typica: Togo (Types, ♂♀, in Museum Berlin).
1895 *Mitrophrys agoma* KARSCH, Entomol. Nachr.Bl., Wien, **21**:355. – Terra typica: Togo (Type, ♀, in Berlin Museum).
1913 *Aegocera trimeni* ssp. *halans* – JORDAN in SEITZ, Großschmetterl. **15**:18.
1913 *Aegocera trimeni* ssp. *halans* forma *transita* JORDAN in SEITZ, Großschmetterl. Erde, **15**:18.
1913 *Aegocera trimeni* ssp. *halans* forma *agoma* JORDAN in SEITZ, Großschmetterl. Erde, **15**:18.

Diagnosis: Fore wings with a pure white discal patch on the underside, at least twice as long as it is broad on both sides of the wing; hind wings white or yellow.

Distribution: Sierra Leone, Togo, Nigeria, Senegal, Sudan, Zaire: Uele, Tshuapa, Sankuru, Ubangi.

Genus **Omphaloceps** HAMPSON

(fig. 18)

1901 *Omphaloceps* HAMPSON, Cat. Lepidopt. Phal. Brit. Mus., **3**:628. – Species typica: *Eusemia triangularis* MABILLE (by original designation).

Diagnosis: Proboscis fully developed; palpi upturned, with the second joint fringed with long hair in front and the third joint long, roughly scaled and slightly porrect; frons with slightly rounded prominence, much narrowed backwards in ♂; antennae distinctly dilated towards extremity; tibiae fringed with hair on the upperside; the hind tibiae and the first tarsal joint in male with a very long tuft of hair; abdomen with dorsal crests except on terminal urites; ♂ with long protrusible tufts of long hair from base. Nervation: in fore wings, vein 2 arising from three quarters of cell; 3, 4, 5 approximated, equidistant; discocellulars nearly straight; 6 from the lower cell-angle; areole rather long and narrow; 7 and stalk of 8 + 9 from its extremity; 10 from its upper margin, near extremity. In hind wings, vein 2 from six sevenths of cell; 3, 4 and 6, 7 from the cell-angles; discocellulars regularly incurved; 5 obsolescent, from middle of discocellulars.

Fig. 18. *Omphaloceps triangularis* (MABILLE). B. M. Noct. 8295

Male genitalia: Uncus about three-fifths of costa in length, narrow, compressed, faintly arched, with dense pilosity. Tubus analis rather narrow, sclerified laterally and ventrally. Fold of vinculum narrow and ending halfway, bearing a bunch of longisch stiff hair. Valva elongate, rather broad, with costa and sacculus nearly parallel, the former incurved, the latter outcurved, especially at middle; apex rounded; fold of sacculus narrow, only reaching to one third of sacculus, ending in a slender harpe angled at middle. Aedeagus about five-sixths of costa in length, moderately robust, swollen in the subproximal portion; vesica bearing numerous cornuti; fultura inferior narrowing distally, with the proximal angles produced and the proximal margin incurved sublaterally; distal angles produced into very narrow processes; fultura superior semicircular, slightly angulated at middle.

Distribution: Most of the Ethiopian region.

Key to the species

1 Fore wings without light spots in basal half . . . . *O. triangularis*, p. 74
1′ Fore wings with two light spots in basal half . . . . *O. daria*, p. 74

Omphaloceps triangularis (MABILLE)

1893 *Eusemia triangularis* MABILLE, Ann. Soc. entomol. Belgique, Bruxelles, **37**:57. – Terra typica: Cameroon (Type, ♂, in British Museum).
1893 *Phasis astrapeus* HOLLAND, Psyche, Cambridge (Massachusetts), **6**:373. – Terra typica: Ogowé (Types, ♂, ♀, in Carnegie Museum).
1901 *Omphaloceps triangularis* – HAMPSON, Cat. Lepidopt. Phal. Brit. Mus., **3**:628–629; fig. 275.
1904 *Rothia mariae* SWINHOE, Trans. r. entomol. Soc. London, **1904**:139. – Terra typica: Kampala, Uganda (Type, ♀, in British Museum).
1913 *Schausia triangularis* – JORDAN in SEITZ, Großschmetterl. Erde, **15**:16; pl. 2d.

Diagnosis: Body and wings dark chocolate brown; base of palpi, extremity of the second palpal joint, the sides of frons, the basal joint of antennae whitish; pectus orange; legs brown with some orange. Fore wings with a narrow, about 2 mm wide, oblique yellow band from below costa just beyond the middle to above tornus; hind wings with an oblique yellow triangular patch from the basal half of costa to near termen on vein 2; costal area whitish. The markings are sometimes whitish (*mariae*). Forms intermediate between *triangularis* and *mariae* occur sometimes (e.g. in Kibali-Ituri, Zaire): fore wing with the band yellow, hind wings with the patch white. Length of fore wing 23–26 mm.

Distribution: Cameroon, Gaboon, Ivory Coast, N. and W. Zaire: Ubangui, Uele, Kibali-Ituri, Kinshasa, Sankuru.

Omphaloceps daria (DRUCE)

1895 *Agarista daria* DRUCE, Ann. Mag. nat. Hist., London, (6) **15**:42. – Terra typica: Lindi, Tanzania (Type, ♀, in British Museum).
1896 *Aegocera triplagiata* ROTHSCHILD, Novit. Zool., **3**:95. – Terra typica: M'pta, Loangwa River, Zambesia (Type, ♂, in British Museum).
1896 *Aegocera triplagiata* var. *dispar* KIRBY, Ann. Mag. nat. Hist., London (6) **18**:377; pl. 19, fig. 1. – Terra typica: East Africa (Type in British Museum).
1898 *Mitrophrys meraca* KARSCH, Entomol Nachr.Bl., Wien, **24**:332. – Terra typica: Southwest Africa (Type, ♂, in Museum Berlin).
1901 *Omphaloceps daria* – HAMPSON, Cat. Lepidopt. Phal. Brit. Mus., **3**:628; pl. 54, fig. 18.
1913 *Schausia daria* – JORDAN in SEITZ, Großschmetterl. Erde, **15**:16; pl. 2d.
1913 *Schausia daria* forma *triplagiata* – JORDAN in SEITZ, Großschmetterl. Erde, **15**:16; pl. 2d.

Diagnosis: Body brownish black; palpi fringed with orange hair; the sides of frons white; collar with white points; thorax with dorsal white stripe; fore tibiae banded with orange; abdomen sometimes orange laterally. Fore wings brownish black; spots white or yellow, viz., an oblique triangular antemedial patch from the subcosta to above vein 1; a band across cell; an oblique postmedial band from below costa to interspace I; silvery white markings, viz., a point below base of costa; an antemedial striga above vein 1; a bar across middle of cell in ♂, and a discocellular bar to dorsum in ♀. Hind wings white, orange yellow or deep orange; terminal half black, narrowing somewhat to anal margin; cilia white at the tips. Two colour phases: *daria* (= *dispar*) with the markings and hind wings white; and *triplagiata* (= *meraca*) with the markings yellow, and the hind wings orange yellow or deep orange. Length of fore wing 24–31 mm.

Distribution: Tanzania, Barotseland, Rhodesia, Republic of South Africa: Namibia; Malawi.

Genus **Epischausia gen. nov.**

(fig. 19)

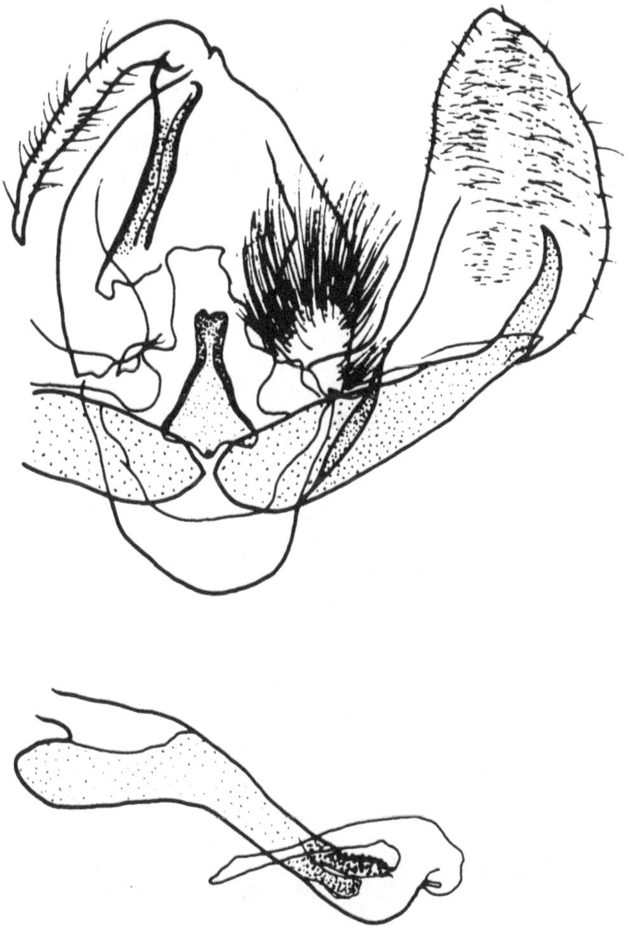

Fig. 19. *Epischausia dispar* (ROTHSCHILD). B. M. Noct. 8578. Typus

S p e c i e s t y p i c a : *Aegocera dispar* ROTHSCHILD (designatione hoc loco).

D i a g n o s i s : Antennae simple, slightly dilated towards the extremity, somewhat less in ♀; palpi with the second joint densely hairy beneath, and with the third joint short, scaly; fore femora covered with long and dense hairs. Nervation: in fore wings, vein 2 arising at about threequarters of the cell; 3, 4, 5 separated, equidistant; discocellulars incurved; 6 from the upper cell-angle; areole small, rhomboidal; 7 and the short stalk of 8 + 9 from its extremity; 10 from its upper edge, near the extremity. In hind wings, vein 2 given off at fourfifths of the cell; 3, 4 from about a point; discocellulars incurved; 5 from middle thereof; 6, 7 from a point; 8 anastomosing for a short while with the cell at the base, then diverging.

Male genitalia: Uncus about as long as two-thirds of costa, compressed, bent at the base, then nearly straight, with the pilosity rather sparse, and a terminal hook. Tubus analis slightly shorter than the uncus, narrow, well sclerified. Fold of vinculum broad proximally, with a fan-shaped long pilosity. Valva with the basal portion rather narrow, broadening and about leaf-shaped distally, the sacculus rather markedly outcurved at middle; fold of sacculus broad, nearly reaching to middle of valva; harpe moderately long, faintly arched. Aedeagus only slightly shorter than costa, with the proximal half spoon-shaped; terminal part rather robust, nearly straight, with a terminal area of minutest cornuti; fultura inferior more or less triangular, with the sides incurved; proximal margin slightly produced at middle; terminal margin faintly broadened, and with a small median excision. Saccus short but broad, rounded.

D i s t r i b u t i o n : East Africa.

Epischausia dispar (ROTHSCHILD)

1896 *Aegocera dispar* ROTHSCHILD, Novit. Zool., **3**:43; pl. 27, 28. – T e r r a t y p i c a : East Africa, Wassein (Types, ♂, ♀, in British Museum).
1901 *Tuerta dispar* – HAMPSON, Cat. Lepidopt. Phal. Brit. Mus., **3**:627.
1913 *Schausia* (?) *dispar* – JORDAN in SEITZ, Großschmetterl. Erde, **15**:17; pl. 4b.
1913 *Schausia flavifrons* BETHUNE-BAKER, Ann. Mag. nat. Hist., London, **12**:66. – T e r r a t y p i c a : Mombasa (Type, ♀, in Museum Oxford University).

D i a g n o s i s : Body black; head, collar and some hair on shoulders orange; antennae and the last palpal joint black; pectus orange, fore coxae and stripes on femora and tibiae orange; abdomen with the anal tuft tipped with orange in ♀. Fore wings black; an oblique yellowish-white wedge-shaped antemedial patch from subcosta to above vein 1, strongly reduced and obsolescent in ♀; an oblique postmedial band between veins 2 and 8, rounded at the extremities; a silvery blue bar in end of cell; a silvery blue discocellular bar, a line beyond the postmedial band, and a streak below vein 2. Hind wings black; ♂ with a yellowish-white wedge-shaped patch in, below and beyond cell; in ♀ without light markings. Length of fore wing 20 mm (♂), 25 mm (♀).

D i s t r i b u t i o n : Coast of Kenya (Wassein); Mombasa.

Genus **Lophonotidia** HAMPSON
(fig. 20)

1901 *Lophonotidia* HAMPSON, Cat. Lepidopt. Phal. Brit. Mus., 3:617. – Species typica: *Lophonotidia nocturna* HAMPSON (by original designation).

Diagnosis: Proboscis fully developed. Antennae moderately dilated towards the extremity. Palpi upturned, reaching the frons and clothed with long hair, the third joint minute, hairy. Frons with a large pointed triangular prominence. Head and thorax clothed with long hair. Tibiae fringed with long hair. Abdomen with large dorsal crests on the three proximal urites; in male, besides that, lateral crests at the base. Termen of fore wings crenulate. Nervation: in fore wings, vein 2 from about two-thirds of cell; 3, 4, 5 approximated and equidistant; 6 from the upper cell-angle; areole narrow; 7 and the moderately long stalk of 8 + 9 from the extremity of the areole; 10 from slightly above it. In hind wings vein 2 given off from five-sixths of cell; 3, 4 and 6, 7 from the cell-angles; discocellulars incurved; termen sinuous and slightly incised between veins 4 and 6; 5 obsolescent from middle of discocellulars.

Male genitalia: Uncus about two-thirds of costa in length, slender, compressed, strongly arched shortly after base, then very faintly bent, bearing a minute terminal hook. Tubus analis short and broad, sclerified ventrally. Fold of vinculum broadly rounded, bearing a fringe of long stiff hair. Valva sickle-shaped, with the costa regularly incurved, and the apex bearing a few stiff bristles; fold of sacculus broad, reaching slightly beyond one-third of the valva, ending in a small sickle-shaped harpe. Aedeagus slightly shorter than the costa, moderately robust, nearly straight, spoon-shaped proximally, bearing terminally a short stripe of spine-like cornuti; fultura inferior narrowing distally, with the proximal margin bearing a pair short median processes, and with the angles somewhat produced. Saccus semi-elliptical.

Distribution: South Africa.

Key to the species

1 Head and thorax dark red-brown; fore wings deep chocolate red-brown; hind wings with the base black . . . . *L. nocturna*, p. 77

1' Head and thorax blackish; fore wings fuscous black; hind wings with basal half white . . *L. melanoleuca*, p. 78

Lophonotidia nocturna HAMPSON

1901 *Lophonotidia nocturna* HAMPSON, Cat. Lepidopt. Phal. Brit. Mus., 3:617–618; fig. 268. – Terra typica: Lake Nyasa, Malawi (Type, ♂, in British Museum).

Diagnosis: Head and thorax dark red-brown, mixed white; antennae sprinkled with blue; pectus in front and flanks, and fore coxae orange; fore legs sprinkled with white; fringes of tibiae mixed white; tufts at the base of abdomen rufous; abdomen black; dorsal tufts red-brown, anal tuft mixed white. Fore wings deep chocolate red-brown; an antemedial bisinuate olive-green line with silvery-blue scales on its proximal side; a silvery-blue spot with some olive round it in cell-middle; a bar across the cell-end, and a discocellular bar with some olive before and beyond it, all silvery-blue; a waved olive line from vein 3 beyond the cell-angle to dorsum; an obsolescent white band irrorated with olive and red-brown, from two-thirds of costa

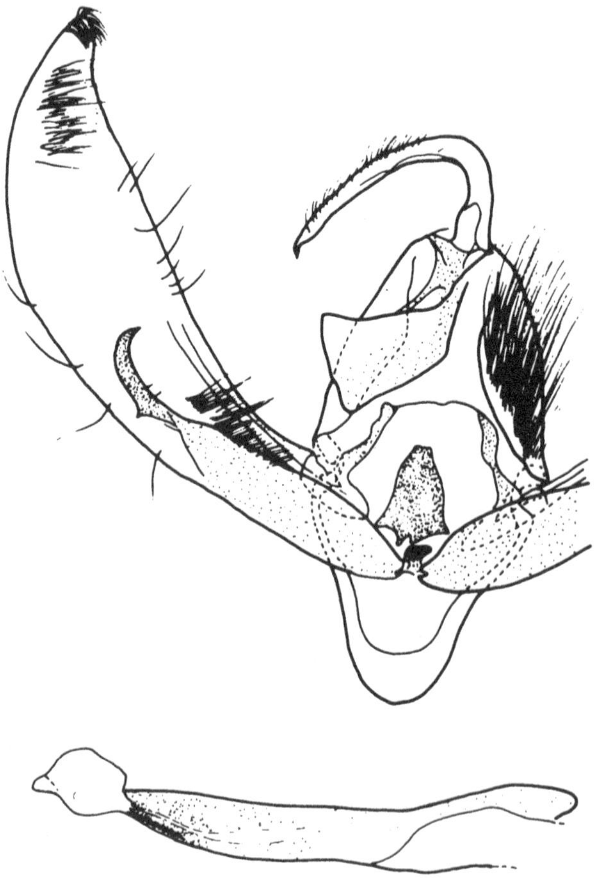

Fig. 20. *Lophonotidia nocturna* HAMPSON. B. M. Noct. 8456. Typus

to tornus; terminal area sprinkled with grey and with some subterminal olive marks and a streak in interspace IV. Hind wings black, with a large white patch clothed with black hair and tinged olive, from the subcostal vein to middle of the anal margin; cilia chequered white and black. Length of fore wing 23 mm.

Distribution: Malawi.

Lophonotidia melanoleuca JANSE

1939 *Lophonotidia melanoleuca* JANSE, Moths South Africa, **3**:41, fig. 14; 2, fig. 10. – Terra typica: Matetsi, Rhodesia (Types, ♂, ♀, in Transvaal Museum).

Diagnosis: Head, body and fore wings fuscous black; head, palpi, dorsum of thorax and legs mixed white hairs; pectus, extremity of abdomen and coxae with yellow pilosity; abdominal crests purplish, shining; anal tuft white in ♂, orange in ♀. Fore wings with a broad, ill-defined oblique white band from just beyond the cell to tornus; some white scales beyond it to the termen from vein 5 to tornus, denser terminally before termen towards apex; antemedial line irregular, zigzagged, olive green proximally, purplish-black distally; some purplish-black scaling in cell beyond it; discocellular mark with a metallic centre and an imperfect olive-green ring, followed by a narrow dark purplish cell-patch and a semi-lunar patch before the white; the so-called reniform mark continued to two-thirds of dorsum as an irregular, dentate, olive-green to yellowish-green scaling preceded by purplish-black; a subterminal series of purplish-black maculae between veins, followed by semi-lunar white scaling at the termen; cilia purplish-black, paler at the extremities, chequered white in ♂. Hind wings white in the basal half, with some long black hairs; a dark fuscous or fuscous-black border, with some pale yellow scaling near the termen on vein 2 and inner folds, very small in ♀; cilia concolorous, chequered between the veins white. Length of fore wing 20 mm in ♂, 22–31 mm in ♀.

D i s t r i b u t i o n : Rhodesia, Transvaal.

Genus **Misa** KARSCH
(fig. 21)

1895 *Misa* KARSCH, Entomol. Nachr.Bl., Wien, **21** :349. – S p e c i e s t y p i c a : *Misa memnonia* KARSCH (by original designation).

D i a g n o s i s : Proboscis fully developed. Antennae somewhat dilated before the extremity which is long-stretched and pointed. Palpi upturned, with the second joint robust, moderately hairy, and the third short. Frons with conical truncate projection. Nervation: Fore wings of ♂ with a costal apparently stridulating apparatus with the resulting deviation of nervures; vein 2 given off from three-quarters of cell; 3, 4, 5 approximated and equidistant; discocellulars incurved; areole very narrow; 6 from the upper cell-angle; 7 and stalk of 8 + 9 from extremity of the areole; 10 from the upper angle thereof. In hind wings, vein 2 branching off very near the lower cell-angle; 3, 4 and 6, 7 from a point at the cell-angles; cell very long.

Male genitalia: Uncus relatively short, arched, with a minute terminal hook. Tubus analis broad, sclerified ventrally, or sometimes at the base. Valva elongate, typically banana-shaped, but sometimes with the costa straight or slightly outcurved in median portion; subapical edge of sacculus and apex with numerous stiff, spinelike hairs; fold of the sacculus narrow, not reaching middle of valva; harpe relatively small, arched. Aedeagus shorter than valva, relatively robust, straight or faintly arched, with a subterminal patch of cornuti; fultura inferior deeply cleft at middle of the proximal margin; distal margin straight. Saccus semi-oval.

D i s t r i b u t i o n : West and parts of Central Africa.

Key to the species

1 Head black . . . . *M. schultzei*, p. 81
1′ Head red or reddish-brown 2
2 Fore wings with a subcostal stripe of raised scales . . . . *M. costistrigata*, p. 81
2′ Fore wings without such stripe 3

3 Fore wings with the white discal band broadening towards dorsum but not reaching tornus; hind wings with the discal band very broad, the black base narrow, and the terminal black band much narrower than the white discal band . . . . . *M. cosmetica,* p. 81
3′ Fore wings with the white discal band narrower, but reaching tornus; hind wings with the discal band narrow, the black base broad, and the terminal black band as wide as the white discal band . . . . *M. memnonia*, p. 80

Misa memnonia KARSCH

1895 *Misa memnonia* KARSCH, Entomol. Nachr.-Bl., Wien, **21**:356; pl. 2, fig. 9. – Terra typica: Togo (Type, ♀, in Berlin Museum).

Diagnosis: Head, collar and thorax deep red-brown in ♂, thorax in ♀ blackish-brown; last palpal joint, frontal prominence and antennae black; metathorax with a red patch; fore coxae red in front; abdomen blackish-brown with the dorsal crests and the three last urites red above; anal tuft in ♀ blackish-brown. Fore wings deep red-brown in ♂, blackish-brown in ♀; a very oblique yellowish-white medial band 2–3 mm wide, ending on dorsum near tornus, with diffused leaden-silvery scales before and beyond it; leaden-silvery markings, viz., a spot below base of costa; a subbasal line from costa to vein 1; a very oblique antemedial line, from costa to the media where it is angled and incurved. Hind wings blackish-brown, with a broad antemedial yellowish-white band; cilia white at the anal angle. Length of fore wing about 20 mm.

Distribution: West Africa (Senegal, Gambia, Ghana, Nigeria, Togo, Cameroon).

Fig. 21. *Misa memnonia* KARSCH. B. M. Noct. 8478

Misa cosmetica KARSCH

1898 *Misa cosmetica* KARSCH, Entomol. Nachr.Bl., Wien, **24**:334. – Terra typica: Misahöhe, Togo (Type, ♂, in Berlin Museum).
1912 *Tuerta cosmetica* – STRAND, Lepidopt. Cat., **5**:29.
1913 *Schausia cosmetica* – JORDAN in SEITZ, Großschmetterl. Erde, **15**:16; pl. 2g.

Diagnosis: Head and collar dark red; thorax and legs black; abdomen dark red above; extremity and underside black. Wings velvety bluish-black; fore wings with a medial white, oblique, rather wide band, from costa to dorsum, slightly before tornus, somewhat outcurved between veins 3 and 4, proximally straight and edged on both sides with metallic blue scales; a similar scale-line between the white band and the base, and a short subbasal line; nervures in distal area finely scaled with metallic blue; costal fold with silvery-white scales. Hind wings with a medial white band, narrow costally, broadening towards the anal margin, irregularly dentate proximally, rounded distally in middle, and broadly incurved in costal and anal areas. Length of fore wing 22,5–27 mm.

Distribution: Togo, South Cameroon; Zaire (Kasai, Lulua, Sankuru, Equateur, Uele).

Misa costistrigata (BETHUNE-BAKER)

1927 *Tuerta costistrigata* BETHUNE-BAKER, Ann. Mag. nat. Hist., London, (9) **20**:323–324. – Terra typica: Bitje, Cameroon (Type, ♂, in British Museum).
1934 *Schausia costistrigata* – GAEDE in SEITZ, Großschmetterl. Erde, **15**:27.

Diagnosis: Head brownish-chestnut-carmine-red; thorax black; femora fringed with long deep-red hair; abdomen black with a dorsal stripe of deep-bronzy-chestnut-red, extending laterally towards the extremity. Wings black; fore wings with a broad oblique white band from costa to dorsum; a subcostal stripe of differently placed scales from the proximal edge of the white band almost into the subapical area, white in white area, but neutral grey in the black area; in basal area, three indistinct lines of scattered blue scales; nervures in distal area with finely bluish scales. Hind wings with a very narrow black basal area, followed by a very broad white area, and a very broad black margin. Length of fore wing about 22 mm.

Distribution: Cameroon.

Species incertae sedis

Misa schultzei (AURIVILLIUS)

1925 *Tuerta schultzei* AURIVILLIUS, Ergebn. 2. Dtsch. Afr. Exp., **1** (18):1311–1312. – Terra typica: Kimuenza, Kongo = Zaire (Type, ♀, in Museum Berlin).

Diagnosis: Body black with white hair or scales mixed; fore and mid coxae and femora with long red-brown pilosity at the bases; metathorax above and the IInd and IIId urites with long paired red-brown tufts. Fore wings black; basal area with some blue scales; veins in distal area sprinkled with light blue; a pure white discal band encroaching into the cell-extremity, nearly straight, about 5 mm wide in interspace V, running from middle of costa and reaching dorsum slightly beyond middle; cilia dark. Hind wings black, with a broad white discal band, reaching the anal margin, but costally only reaching vein 8, regularly outcurved; cilia dark. Length of fore wing 27 mm. – This species is unknown to the writer.

Distribution: Zaire (Type only known).

Genus **Schausia** KARSCH
(fig. 22)

1895 *Schausia* KARSCH, Entomol. Nachr.Bl., Wien, **21**:346. – Species typica: *Aegocera leona* SCHAUS (by original designation).

Diagnosis: Proboscis fully developed; palpi upturned, fringed with long hair in front, the third joint short, clothed with long rough hair; frons with a truncate conical projection with raised rim at extremity, in male very narrow, especially posteriorly; antennae strongly dilated into a knob at the extremity; tibiae fringed with long hair on outer side, without spines; hind tibiae with long erect pilosity; abdomen with dorsal crests on basal urites. Nervation: In fore wings, vein 2 given off at about three-quarters of cell; 3, 4, 5 approximated and equidistant; discocellulars incurved; areole very narrow; 6 from the areole near the base; 7, 8 + 9 and 10 stalked, from the extremity of the areole. In hind wings, vein 2 given off slightly beyond five sevenths of cell; 3, 4 much approximated; discocellulars well incurved; 5 obsolescent, from middle of discocellulars; 6, 7 from the upper cell-angle.

Male genitalia: Uncus about half the costa in length, narrow, arched, with a minute terminal hook. Tubus analis broad, sclerified ventrally. Fold of vinculum narrow, with a submarginal row of moderately long, stiff hairs. Valva elongate, moderately broad; costa nearly straight; apex subacute; sacculus nearly parallel with costa to slightly beyond middle, then outcurved and faintly arched, so that the distal half of valva is broader than the proximal half; fold of sacculus moderately broad, reaching to base of the broad portion of valva, and ending in a short spine-like harpe, directed transversally towards costa. Aedeagus about as long as costa, slender, boomerang-shaped; fultura inferior broader proximally, with the proximal margin truncate-angulate, very faintly notched at middle; lateral margins incurved. Saccus elongate semi-elliptical.

Distribution: Most of tropical Africa.

Key to the species

1 Hind wings yellowish-white . . . . *S. leona*, p. 82
1′ Hind wings orange or red 2
2 Hind wings crimson . . . . *S. greenawayi*, p. 85
2′ Hind wings orange 3
3 Fore wings with a white wedge-shaped fascia from base . . . . *S. gladiatoria*, p. 84
3′ Fore wings with a medial and a postmedial white bands . . . . *S. coryndoni*, p. 84

Schausia leona (SCHAUS)

1893 *Aegocera leona* SCHAUS, Lepidopt. Sierra Leone:20; pl. 1, fig. 1. – Terra typica: Sierra Leone (Type, ♂, in U.S. National Museum).
1893 *Schausia leona* – KARSCH, Entomol. Nachr.Bl., Wien, **21**:346.

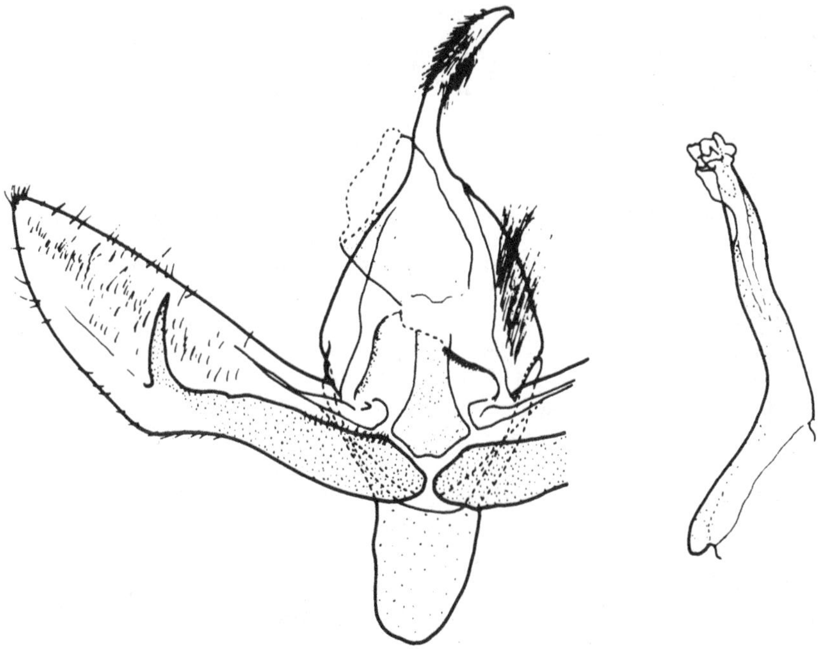

Fig. 22. *Schausia leona* (SCHAUS). B. M. Noct. 8431

Diagnosis: Body and fore wings deep chocolate-brown; palpi with fulvous hair-fringe; coxae with fulvous pilosity. Fore wings with a very oblique, diamond-shaped silvery-white patch from costa to middle of the submedian fold; a large lunulate silvery-white patch beyond cell between veins 3 and 8; purplish silvery-brown markings, viz., a spot on middle of dorsum; an oblique line across the cell-extremity to near tornus, where it is dilated into a large patch; a discocellular bar; a subterminal series of lunulate spots; and a terminal band. Hind wings yellowish-white, with a very broad black-brown terminal band, continued on costa to base; termen tinged with chocolate-brown. Length of fore wing 19–24 mm.

Distribution: West and Central Africa.

Schausia leona leona (SCHAUS)

1913 *Schausia leona* ssp. *leona* – JORDAN in SEITZ, Großschmetterl. Erde, **15**:15; pl. 2f.

Diagnosis: Hind wings with the terminal band twice irregularly bent, and not expanding on the anal margin.

Distribution: Sierra Leone, Gold Coast, Ivory Coast to Niger region.

Schausia leona congoana JORDAN

1908 *Schausia leona* ssp. *congoana* JORDAN, Novit. Zool., **15**:253. – Terra typica: Luebo, Kasai Province, Congo = Zaire (Type, ♂, in British Museum).

D i a g n o s i s : Hind wings with terminal band less bent, narrowing towards the anal angle; anal margin black.

D i s t r i b u t i o n : Congo-Brazzaville, Zaire (Uele, Equateur, Kasai, Sankuru).

Schausia gladiatoria (HOLLAND)

1893 *Aegocera gladiatoria* HOLLAND, Psyche, Cambridge (Massachusetts), **6**:373. – T e r r a t y p i c a : Ogowé River, West Africa (Types, ♂, ♀, in Carnegie Museum).
1901 *Schausia gladiatoria* – HAMPSON, Cat. Lepidopt. Phal. Brit. Mus., **3**:616; pl. 54, fig. 8.

D i a g n o s i s : Head and thorax red-brown; extremity of the second, and the third palpal joint, frons, vertex, collar and a stripe on thorax whitish; pectus and legs orange; fore femora fringed with whitish; fore and mid tibiae and tarsi banded with rufous; abdomen orange with a rufous dorsal stripe. Fore wings chocolate-brown; a creamy-white fascia from base of costa to beyond middle in the submedian interspace, ending in a point; a broad slightly oblique whitish postmedial band from vein 2 to below costa, pointed at the extremity; leaden-silvery markings, viz., an oblique bar in cell-end; a discocellular bar; an oblique line from the submedian fascia before middle to dorsum, along which it runs to near tornus, where it is joined by an oblique line from near base of vein 2; an oblique subterminal line from costa to vein 2; a terminal line. Hind wings reddish-orange, with a terminal red-brown band, narrowing to tornus, with an orange spot on termen in interspace I. Length of fore wing 15–18 mm (♂), 17–18 mm (♀).

D i s t r i b u t i o n : Congo-Brazzaville, Gaboon, Ivory Coast, Sierra Leone, Gold Coast, Sénégal, Zaire (Kinshasa, Kasai).

Schausia coryndoni (ROTHSCHILD)

1896 *Aegocera coryndoni* ROTHSCHILD, Novit. Zool., **3**:96. – T e r r a t y p i c a : Loangwa River, Barotseland (Type, ♀, in Rothschild collection, Tring; apparently lost).
1896 *Aegoceropsis separata* WEYMER, Berliner entomol. Z., **1896**:87. – T e r r a t y p i c a : Tanganyika = Tanzania (Type, ♀ [fide STRAND] in Berlin Museum).
1913 *Schausia coryndoni* – JORDAN in SEITZ, Großschmetterl. Erde, **15**:16.

D i a g n o s i s : Head yellow spotted with black; antennae black, with white scales above; palpi yellow, the first and the second joints spotted with black, the third nearly all black; collar yellow spotted with black; tegulae mixed whitish and chocolate-brown hair, with a black shoulder-spot; mesothorax grey; metathorax, pextus and legs orange; fore coxae grey behind; tarsi orange ringed with black; abdomen orange; dorsal crests, anal tuft and ventral bands on terminal urites black. Fore wings chocolate-brown, basal and terminal areas tinged with grey, disc with ochreous; a medial white band from below costa to submedian fold, expanding below; a similar band beyond cell between veins 3 and 8, each patch with an olive line round it with silvery scales at its edge; silvery spots in middle of cell above and below the median band, and a discocellular bar; cilia grey. Hind wings orange; a terminal black band, narrowing towards the anal angle; cilia mixed white and black. Length of fore wing 25 mm.

D i s t r i b u t i o n : South Africa: Barotseland, Transvaal; Tanzania.

Schausia greenawayi STONEHAM

1963 *Schausia greenawayi* STONEHAM, Bull. Stoneham Mus. Nairobi, **80**:2. – Terra typica: Busia, Uganda (Type, sex not stated, in Stoneham Museum).

Diagnosis: Fore wings black with white fringes to discal margins, divided by an oblique pale yellow transversal line from the centre of the costa to the anal angle. The costa is thinly pale yellow from the middle to the base, where it is bifurcated by short black line. The hindwings are crimson with broad black borders. – This species is unknown to the writer and is here considered as *incertae sedis.*

Distribution: Uganda.

Nervation: in fore wings, vein 2 given off at about three-quarters of cell; 3, 4, 5 well separated, equidistant, upper discocellular incurved; areole small and narrow; 6 from distinctly below the lower cell-angle; 7 and 8 + 9 from the extremity of the areole; 10 from the upper side of the areole, near its extremity. In hind wings, vein 2 given off at two-thirds of cell; 3, 4 and 6, 7 stalked; discocellulars nearly straight; 5 obsolescent, from middle of the discocellulars.

Nervation: in fore wings, vein 2 given off at two-thirds of cell; 3, 4, 5 well separated, equidistant; discocellulars incurved, 6 from the lower cell-angle; areole very narrow; 7 and the short stalk of 8 + 9 from the extremity of the areole; 10 slightly above. In hind wings, vein 2 given off from three-quarters of cell; 3, 4 and 6, 7 from the cell-angles; discocellulars incurved; 5 obsolescent, from middle of the discocellulars; 8 anastomosing at a point with the cell.

Genus **Hoplarista** HAMPSON

(fig. 23)

1910 *Hoplarista* HAMPSON, Proc. Zool. Soc. London, **1910**:399. – Species typica: *Hoplarista haemaplaga* HAMPSON (by original designation).
1913 *Paida* JORDAN (partim) in SEITZ, Großschmetterl. Erde, **15**:13.

Diagnosis: Proboscis fully developed. Antennae slightly dilated towards the extremity. Second palpal joint obliquely upturned and moderately fringed with hair; third palpal joint rather long, porrect, and faintly dilated at the extremity. Frons with a long flattened corneous process, bifid at the extremity, with a corneous plate below it, excised in front. Metathorax with a spreading crest. Tibiae moderately fringed with hair. Abdomen with a basal hairy crest.

Male genitalia: Uncus about one-half of costa in length, slender, nearly straight, pointed. Tubus analis short and broad, well sclerified dorsally, and a little ventrally. Vinculum very broad and with the sides fused except basally; fold very narrow, without fringe of stiff hair. Valva leaf-shaped, with the apex broadly rounded; fold of sacculus very narrow, ending in a very short and small arched harpe, reaching to two-thirds of the valva. Aedeagus very slightly shorter than than the valva, moderately robust, more or less straight; terminal portion bearing dorsally a number of minute cornuti; fultura inferior more or less sand-clock-shaped, faintly three-pronged proximally, and excised at the middle distally. Saccus semi-circular.

Distribution: Zambia.

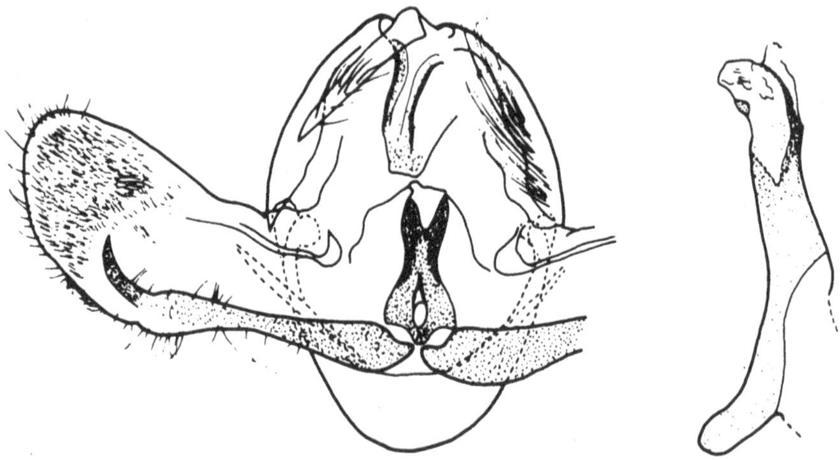

Fig. 23. *Hoplarista haemaplaga* HAMPSON. B. M. Noct. 8461. Typus

Hoplarista haemaplaga HAMPSON

1910 *Hoplarista haemaplaga* HAMPSON, Proc. zool. Soc. London, **1910**:399; pl. 36, fig. 24. – Terra typica: Lake Bangweolo, Zambia (Types, ♂, ♀, in British Museum).
1913 *Paida haemaplaga* – JORDAN in SEITZ, Großschmetterl. Erde **15**:13; pl. 4a.

Diagnosis: Head and thorax dark chocolate brown, mixed with yellowish scales; frons spotted laterally with yellowish white; pectus orange; lags banded with orange and black above; abdomen black banded wiht orange above, dark brown banded with yellow below. Fore wings dark chocolate brown, sprinkled with silvery blue; a triangular yellow antemedial cell-patch, with a yellow spot in it in female; an elongate Indian red patch below it, connected with the oblique yellow medial band extending from below costa to the submedian fold, narrower in female, edged laterally with Indian red; a postmedial yellow patch from below costa to vein 2 in male and to vein 4 in female. Hind wings orange, with a slight black discocellular lunule; a blackish brown terminal band, broader in female, bearing in male a small yellow spot on termen in the submedian fold. Length of fore wing 23–24 mm.

Distribution: Zambia.

Genus **Paida** JORDAN
(fig. 24)

1896 *Paida* JORDAN, Novit. Zool., 3:47. – Species typica: *Pais pulchra* TRIMEN (by original designation).

Diagnosis: Palpi very long, porrect, with the second joint fringed with short hair, and the third nearly as long as the second, smoothly scaled. Antennae almost simple, distinctly dilated towards the extremity. Frons with a trifurcate or conical prominence. Hind tibiae fringed with long rough hair. Abdomen of male with long lateral tufts of hair on the terminal segments. Nervation: in fore wings, vein 2 issued from two-thirds of cell; cell long, about two-thirds of the wing; veins 3 and 4 widely separated; 4 and 5 from a point; discocellulars well incurved; areole minute; 6 and stalk of 7 and 8 + 9 from the extremity of areole; 10 from upper margin of areole. In hind wings, vein 2 given off from five-sevenths of cell; 3, 4 and 6, 7 at a point from the cell-angles; discocellulars faintly angled inwards at the base of vein 5 which is obsolescent; in male inner margin and anal angle fringed with very long hair; there also are some long hairs on the veins of the underside of both wings.

Male genitalia: Uncus relatively small, compressed, faintly bent, pointed. Tubus analis rather broad, sclerified subdorsally. Vincula very broad, fused medially, with the fold rather narrow, and with hair-fringe at the bend and along the margin. Valva moderately broad, upcurved, and the sacculus more distinctly outcurved beyond the middle, otherwise more or less straight; apex rounded; fold of sacculus half as broad as the basal portion of the valva, but soon narrowing, reaching to slightly more than one-third of the sacculus, ending in a short, slender, nearly straight harpe. Aedeagus about four-fifths of costa in length, robust, with the basal spoon-shaped portion slightly shorter than the straight distal part, and placed at an angle of about 120°; terminal half of the aedeagus with a number of minutest warts and a lateral lobe-shaped patch of minute spines; a few terminal denticulations dorsally; vesica forked and irregularly shaped; fultura inferior broad, narrowing distally, with the proximal margin excised at middle and laterally; distal margin deeply bilobate. Saccus short, elliptical.

Distribution: South and South-East Africa.

Paida pulchra (TRIMEN)

1863 *Pais pulchra* TRIMEN, Trans. r. entomol. Soc. London, (3) **1**:524. – Terra typica: Namibia (Type, ♀, in ?).
1879 *Pais gordoni* BUTLER, Entomologist's mon. Mag., **16**:10. – Terra typica: Tugela River, Natal (Type, ♂, in British Museum).
1896 *Paida pulchra* – JORDAN, Novit. Zool., **3**:4.

Diagnosis: Head and thorax above white; palpi above and whole of the last joint, antennae and markings of the head black; collar striped with black; tegulae edged with black, with orange hair on the upper edge; thorax striped with black; pectus and legs with white and orange pilosity; legs barred with black; abdomen orange with black edges to the rings and lateral black points. Fore wings yellowish white; costal and dorsal areas sprinkled with black; an Indian red streak across the dorsal area to the outer line; subbasal and inner lines black, the latter double, interrupted on the red streak; in and below cell, ocellate spots with bluish and black centres and black edges; beyond cell, a third similar ocellate spot; outer line black, beyond three-quarters of costa to near tornus, followed by an Indian red line; terminal area with black internervural streaks; cilia black, tipped with white. Hind wings in male reddish orange, redder terminally; discocellualrs with a dark lunule; red terminal area spotted with orange; cilia black, tipped with yellow; long tornal hair yellowish; hind wings in female orange; a black discocellular bar; an undulate

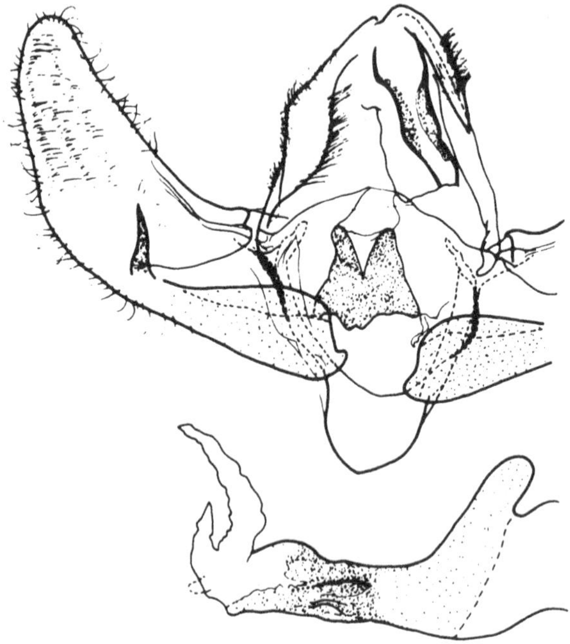

Fig. 24. *Paida pulchra* (TRIMEN). B. M. Noct. 8427

subterminal line connected with the termen by streaks in the interspaces; cilia black, tipped with orange. Length of fore wing 18–22 mm.

Distribution: Republic of South Africa: Natal, Transvaal, Namibia; Rhodesia.

Genus **Pseudopais** BARTEL

(fig. 25)

1903 *Pseudopais* BARTEL, Verh. zool.-bot. Ges. Wien, **53**:127. – Species typica: *Pseudopais nigrobasalis* BARTEL (by original designation).

Diagnosis: Proboscis fully developed; antennae with a well marked subterminal swelling and a pointed extremity; palpi short, appressed, not reaching beyond the frons; pilosity rather long, especially on the first and the underside of the second joint; the third joint pointed, not longer than the second; frons with a well marked truncate conical prominence, with raised rim at the extremity; beneath, a more or less cylindrical process; abdomen in ♂ with thick lateral tufts of hair behind the stigmata. Wings rather elongate and narrow, with the termen well oblique; fore wing with the cell long; hind wing with the cell rather long. Nervation: in fore wings, vein 2 from three-quarters of cell; 3, 4, 5 separate, equidistant; discocellulars incurved; 6 from the lower cell-angle; areole narrow; 7 and the long stalk of 8 + 9 from its lower angle; 10 from its upper angle. In hind wings, vein 2 given off slightly beyond two-thirds of cell; 3 and 4 slightly separated; 5 obsolescent, from middle of disco-cellulars; 6, 7 from a point in ♂, short stalked in ♀.

Male genitalia: Uncus slightly shorter than costa, compressed, evenly arched, ending in a small hook. Tubus analis short, narrow, well sclerified. Fold of vinculum short but broad, fringed on the inner side with rather long stiff hair, expanding to the distal part of vinculum. Valva moderately broad in the proximal half, then expanding and taking an oval shape; costa incurved for two-thirds of its length, then rounded together with the broadly excurved termen; sacculus faintly incurved beyond the basal third, then outcurved; fold rather broad, reaching to middle of valva and ending in a very short, faintly arched harpe. Aedeagus as long as the valva, rather slender, straight; vesica with a pair of divergent basal patches of cornuti; fultura inferior very deeply excised at middle, so that it consists of a narrow base and two nearly parallel lateral prongs. Saccus broad and short.

Distribution: East Africa.

Pseudopais nigrobasalis BARTEL

1903 *Pseudopais nigrobasalis* BARTEL, Verh. zool.-bot. Ges. Wien, **53**:127. – Terra typica: Lake Nyassa, Unyika (Type, ♀, in Museum Berlin).
1911 *Rothia subterminalis* HAMPSON, Ann. Mag. nat. Hist., London, (8) **8**:414. – Terra typica: Mount Rungwe, German East Africa = Tanzania (Type, ♀, in British Museum).
1913 *Pais nigrobasalis* – JORDAN in SEITZ, Großschmetterl. Erde, **15**:14; pl. 2a.
1920 *Rothia nigrobasalis* – HAMPSON, Cat. Lepidopt. Phal Brit. Mus., Suppl. **2**:564.

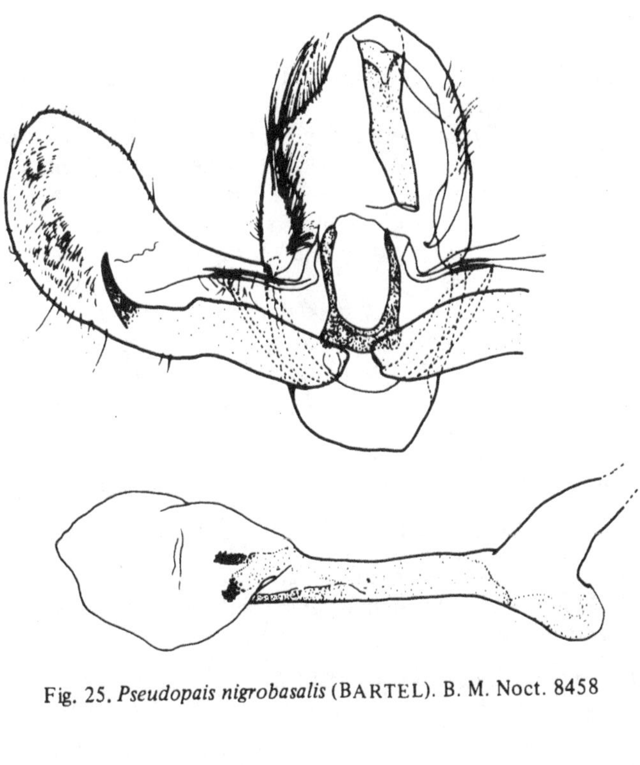

Fig. 25. *Pseudopais nigrobasalis* (BARTEL). B. M. Noct. 8458

D i a g n o s i s : Body and fore wings black; basal joint of antennae white in front; palpi, the sides of frons, collar, shoulders, tegulae, meso- and metathorax with white points; tibiae banded with orange; tarsi ringed with white; abdomen with white segmental lines and sublateral and ventral series of white points; anal tuft black in ♂, orange in ♀. Fore wings with three white basal points, and pale yellow markings, viz., an antemedial series of four small spots; a medial patch from below costa to below cell; a spot above the media; a patch from below costa across cell extremity; a triangular spot below cell extremity; a postmedial band from below costa to vein 3; a subterminal series of greenish-white internervural spots, those towards apex elongate, that above tornus pale yellow; a slight greenish-white streak across middle of dorsum; silver-blue markings, viz., in cell antemedially and submedially; a bar across cell-middle; a patch below cell; a discocellular bar from just below costa to below cell-angle; and a spot above dorsum; cilia pale yellow terminally. Hind wings scarlet with some black at base, with a black terminal band slightly incurved at veins 3 and 2; cilia yellowish white terminally. Length of fore wing 19 mm.

D i s t r i b u t i o n : Tanzania, Malawi.

Genus **Brephos** HÜBNER

(fig. 26)

1813 *Brephos* HÜBNER, Sammlung exot. Schmetterl., 1:pl. 198. – S p e c i e s t y p i c a : *Brephos julia* HÜBNER = *Noctua decora* LINNAEUS (by original designation).
1821 *Pais* HÜBNER, Index exoticorum lepidopterorum (4). – S p e c i e s t y p i c a : *Noctua decora* LINNAEUS (by original designation).

D i a g n o s i s : Proboscis fully developed. Palpi long, reaching beyond the frons, second joint fringed with long hair in front, third joint short, porrect, clothed with rough scales and hairs beneath. Antennae short, distinctly dilated, then tapering to the extremity. Frons with a truncate prominence, ridges on the apical surface slightly lengthened ventrally; forepart of frons surelevated beneath the prominence, with the fore margin of the elevated portion well marked. Tibiae slightly fringed with hair above. Nervation: in fore wings, vein 2 arising from nearly two-thirds of cell; 3, 4, 5 separated, equidistant; discocellulars incurved; 6 from the lower cell-angle; areole narrow; 7 and the long stalk of 8 + 9 from the lower angle of the areole; 10 from the upper angle thereof. In hind wings, vein 2 arising from slightly beyond two-thirds of cell; 3, 4 and 6, 7 stalked, from the cell-angles; discocellulars incurved; 5 obsolescent, from middle of discocellulars.

Male genitalia: Uncus about one-half of costa in length, narrow, arched at the base, then slightly concave, and again arched at the extremity. Tubus analis short, narrow, well sclerified. Vincula broad, with the fold narrow, clothed with short hair. Valva broad, leaf-shaped; termen broader than the basal part; fold of sacculus broad proximally, narrowing distalwards to reach beyond four-fifths of the valva, ending in a short, arched harpe. Aedeagus slightly longer than the valva, robust, faintly arched, with a subterminal patch of spinelike cornuti; vesica bearing a few cornuti; fultura inferior broader basally, with the proximal margin incised at middle and produced at the angles; sides incurved; distal margin incurved, and with the angles produced. Saccus broad, semi-elliptical.

D i s t r i b u t i o n : Southern parts of Africa.

Key to the species

1 Fore wings rusty coloured, with pale yellowish spots and bands, edged with black 2
1′ Fore wings rusty coloured, without pale yellowish markings, but with black, pale-centered spots in cell and on discocellulars 3
2 Fore wings with the pale spots small, especially the two post-discal spots; hind wings with a subterminal series of pale spotlets . . . . ***B. decora***, p. 91
2′ Fore wings with the pale spots large, especially the two post-discal spots; hind wings without pale subterminal dots, and with a large pale, rufous or yellowish discal spot . . . . . ***B. festiva***, p. 92
3 Fore wings with terminal area black with a series of pale yellowish dots . ***B. nyassana***, p. 93
3′ For wings with terminal area concolorous, with dark veins, and without pale dots . . . . ***B. ansorgei***, p. 93

Brephos decora (LINNAEUS)

1762 *Noctua decora* LINNAEUS, Mus. Ulr. :382. – Terra typica: Cape (Type not designated).
1775 *Noctua julia* CRAMER, Pap. exot., **1** :pl. 7, fig. E and F. – Terra typica: Cape? (Type not designated).
1813 ***Brephos julia*** – HÜBNER, Sammlung exot. Schmetterl., **1** :pl. 198.
1821 ***Pais decora*** – HÜBNER, Index exoticorum Lepidopterorum, (**4**).
1881 ***Pais moldaenkei*** DEWITZ, Nova Acta Acad. Leop. Carol. 42:65; pl. 2, fig. 15. – Terra typica: Cape Colony (Type, in ?Museum Berlin).

Diagnosis: Head and thorax black; palpi greyish white; second palpal joint, sides of frons, chin, edges of collar, tegulae, fore coxae in front, tibiae and rings on tarsi, ochreous; femora with yellow pilosity below; abdomen orange, with black bands and dorsal spots; underside black with greyish white bands. Fore wings ochreous,

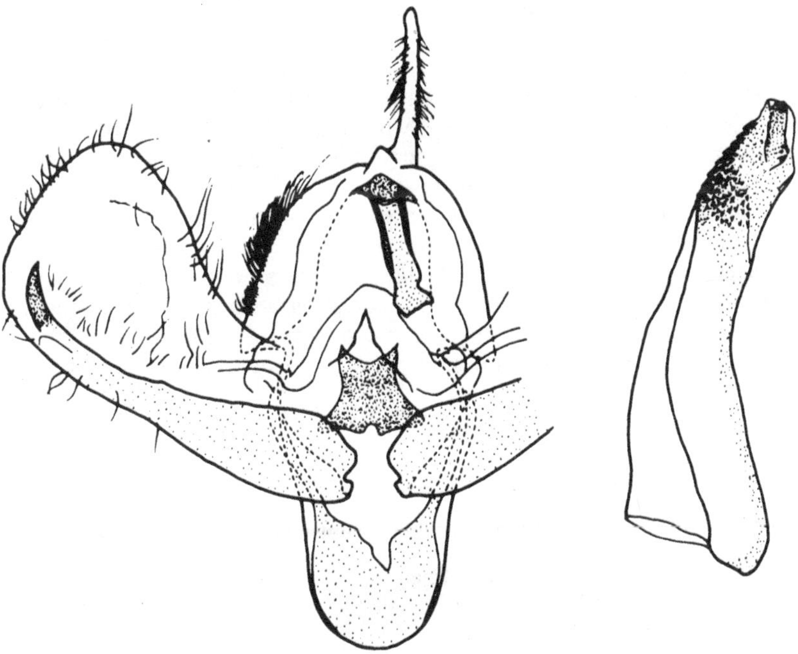

Fig. 26. *Brephos decora* (LINNÉ). B. M. Noct. 8457

more or less sprinkled with black, and tinged with olive in the dorsal area; red markings extensive, occupying more place than the pale yellow ground colour; costa and dorsum black; two oblique black subbasal lines, distal one ending in an indian red, black edged spot below cell; antemedian black line incurved in cell, strongly excurved below it; an indian red, black edged eight-shaped medial spot in and below cell; a black-edged elliptical discocellular spot with red centre, and with a red streak from its lower extremity to the subterminal band; postmedian black line forming inwards a tooth to the discocellular spot at vein 5, below vein 5 retracted to the lower edge of the said spot, then oblique and wavy to dorsum; a sinuous indian red subterminal band, dentate inwards on vein 5, and connected with the black terminal line by streaks on veins, forming a series of yellowish terminal spots; cilia ochreous, spotted with black. Hind wings indian red with a series of terminal yellowish spots; cilia whitish, with black base and spots. Length of fore wing 20--24 mm.

Distribution: Republic of South Africa: Cape Province, Natal.

Brephos festiva (JORDAN)

1913 *Pais festiva* JORDAN in SEITZ, Großschmetterl. Erde, **15**:14; pl. 3f (as *decora*). – Terra typica: Natal (Type, ♂, in British Museum).

Diagnosis: Head and thorax black; palpi mostly blackish, or with variable pale spots; sides of frons, edges of collar and tegulae yellow; chin yellow or black; fore coxae yellow or black; mid and hindcoxae yellow; abdomen orange with black bands and dorsal spots; underside black with pale, sometimes reduced, bands. Fore wings creamy yellow with little or no black sprinkling; indian red markings less extensive than in *B. decora*; costa and dorsum black; two oblique black subbasal lines, the distal one ending in an indian red spot below cell, edged with black; an antemedial black line; an indian red, black-edged, eight-shaped medial spot in and below cell, with one or two yellowish centres to each division; a black elliptical discocellular spot with indian red centre, with small yellowish spots above and below it, and with an indian red streak from its lower extremity to the subterminal band; a postmedial black line forming inwards a tooth to the discocellular mark at vein 5, retracted below vein 3 to the lower edge of the said spot, then oblique and wavy to dorsum; a sinuous indian red subterminal band, dentate inwards on vein 5 and connected with the black terminal line by streaks on veins, forming a series of yellowish terminal spots; cilia ochreous spotted with black. Hind wings indian red with a yellowish ill defined fabiform patch beyond cell, and a small cell-spot; cilia yellowish, with black basal line and spots. Length of fore wing 20–24 mm.

Distribution: Republic of South Africa: Cape Province, Natal, Transvaal.

Brephos festiva festiva (JORDAN)

1913 *Pais festiva* ssp. *festiva* – JORDAN in SEITZ, Großschmetterl. Erde, **15**:14.

Diagnosis: Palpi with well-defined pale spots; fore coxae and extremities of mid and hind coxae with yellow pilosity; pale bands on the underside of the abdomen complete.

Distribution: Natal and Transvaal.

B r e p h o s f e s t i v a n i g r i p e c t u s (JORDAN)

1913 *Pais festiva* ssp. *nigripectus* JORDAN in SEITZ, Großschmetterl. Erde, **15**:14. – T e r r a t y p i c a : Transkei (Type, ♂, in British Museum).

D i a g n o s i s : Palpi with reduced or absent pale spots; pectus and coxae black; pale bands on the underside of abdomen shortened.

D i s t r i b u t i o n : Transkei; Cape Province.

Brephos nyassana (BARTEL)

1903 *Pais nyassana* BARTEL, Verh. zool.-bot. Ges. Wien, **53**:128. – T e r r a t y p i c a : Lake Nyassa (Types, ♂, ♀, in Museum Berlin).
1900 *Pais haplocraspis* HULSTAERT, Rev. Zool. Bot. Afr., Bruxelles, **11**:202. – T e r r a t y p i c a : Elisabethville = Lubumbashi, Shaba (Type, ♀, in Tervuren Museum).

D i a g n o s i s : Head and thorax black; first and second palpal joints, sides of frons, vertex, collar, tegulae spotted with white; tegulae with fringes of orange and white; abdomen black, banded with orange above, and with yellowish white below. Fore wings pale orange rufous; basal area below cell, discal fold and post-medial area tinged with carmine; some black at the base, followed by a black cell-spot; antemedial line interrupted in the submedian fold, then double, filled in with pinkish; a round black spot, centred with grey, in cell; discocellular marking fabiform, filled in with grey, and cut in two by the pinkish red tinge of the discal fold and postmedial line double, filled in with pinkish; terminal area black with a series of more or less triangular yellowish spots; cilia yellowish, spotted with black. Hind wings pale-orange rufous; an obsolete black discocellular lunule and a sinuous postmedial line; termen black, with yellow nervural dots; cilia black at base, then yellowish, spotted with black. Length of fore wing 20 mm.

R e m a r k : *Pais haplocraspis* Hulstaert has the cilia wholly black, and might represent a distinct subspecies but for the fact that "typical" specimens occur both eastwards (Malawi) and westwards (Angola) of its area of distribution.

D i s t r i b u t i o n : Malawi; Angola; Southern Zaire (Shaba).

Brephos ansorgei (JORDAN)

1904 *Pais ansorgei* JORDAN, Novit. Zool., **11**:445. – T e r r a t y p i c a : Bula Matenga, Angola (Type, ♂, in British Museum).
1913 *Paida ansorgei* – JORDAN in SEITZ, Großschmetterl. Erde, **15**:12, pl. 4a.

D i a g n o s i s : Head orange; extremity of the second and whole of the third palpal joint, frontal prominence, and antennae black; thorax black marked with yellow; pectus and legs orange yellow, the latter marked with black; abdomen orange yellow with black bands. Fore wings light indian red; base of the costal area black with a pale spot followed by some silvery blue scales; medial part of costa suffused with black; vein 1 streaked with black, and with two antemedial bars from it to dorsum; a black bar across the middle of ceel, with silvery blue spots on it in cell and on the median vein; discocellulars with a black fabiform spot, centred with silvery blue; postmedial line black with a silvery blue line just beyond it, and with black streaks on the veins from it to the black terminal line; cilia black, becoming brownish at the extremities. Hind wings light indian red; a blackish streak on the basal half of vein 1; a small elliptical black discocellular spot; a curved black subterminal line, with the veins streaked with black between the line and the terminal line; cilia black, with ochreous white extremities. Length of fore wing 23 mm.

Distribution: Angola.

Genus **Crameria** HÜBNER

(fig. 27)

1819 *Crameria* HÜBNER, Verz. bek. Schmetterl., (11):168. – Species typica: *Noctua amabilis* DRURY (designated by HAMPSON 1920).
1854 *Charilina* WALKER, List Lepidopt. Het. Brit. Mus., **2**:560. – Species typica: *Noctua amabilis* DRURY (by monotypy).

Diagnosis: Proboscis fully developed. Antennae almost simple, slightly dilated towards the extremity. Palpi obliquely upturned; second joint fringed with hair becoming long at the extremity, third joint long, slender, naked, porrect, slightly dilated distally, and with the extremity pointed. All tibiae in male, and only the fore tibiae in female fringed with long hair above. Frontal prominence much narrower in male.

Male genitalia: Uncus relatively very small, narrowing distally, and slightly bent terminally. Tubus analis broad. Fold of the vinculum narrow, with a fringe of hair. Valva rather broad, especially basally, with costa and sacculus nearly parallel, the

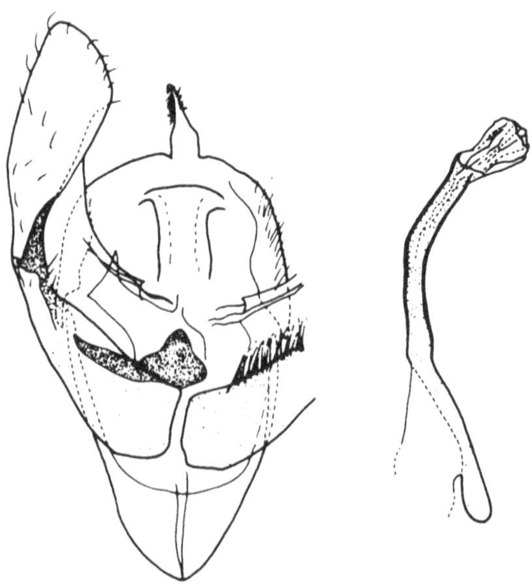

Fig. 27. *Crameria amabilis* (DRURY). B. M. Noct. 8470

former strongly incurved, the latter correspondingly outcurved; termen broad, very faintly outcurved; fold of sacculus about half as broad as the valva at the base, narrowing distally and ending half way on the valva as a short, horn-shaped harpe; valvula basally with a broad process, cut straight terminally, and bearing costally a terminal tooth. Aedeagus about three quarters of valva in length, slender, arched; fultura inferior more or less heart-shaped, bearing a horn-shaped process on the proximal margin. Saccus semi-elliptical.

Distribution: Africa.

Crameria amabilis (DRURY)

1773 *Noctua amabilis* DRURY, Illustr. exot. Entomology, **2**:pl. 13, fig. 3. – Terra typica: "Africa" (Type not designated).
1794 *Noctua alienata* FABRICIUS, Syst. entomol., (3) **2**:60. – Terra typica: Africa (Type not designated).
1819 *Crameria amabilis* – HÜBNER, Verz. bek. Schmetterl., (11):168.
1836 *Aegocera amabilis* – BOISDUVAL, Spéc. gén. Lépidopt., **1**:pl. 14, fig. 6.
1854 *Charilina amabilis* – WALKER, List Lepidopt. Het. Brit. Mus., **2**:56.
1874 *Charilina intercis* FELDER, Reise Novara:pl. 107, fig. 17. – Terra typica: East Africa (Type, in British Museum).
1912 *Charilina amabilis* aberr. *abyssinica* STRAND, Lepidopt. Cat., **5**:21 (= "aberr. 1" HAMPSON, Cat. Lepidopt. Phal. Brit. Mus., **3**:588). – Terra typica: Ethiopia (Type, ♂, in British Museum).
1912 *Charilina amabilis* aberr. *nyassica* STRAND, Lepidopt. Cat., **5**:21 (= "aberr. 2" HAMPSON, Cat. Lepidopt. Phal. Brit. Mus., **3**:588). – Terra typica: Malawi (Type, ♂, in British Museum).
1912 *Charilina amabilis* aberr. *accra* STRAND, Lepidopt. Cat., **5**:21 (= "aberr. 3" HAMPSON, Cat. Lepidopt. Phal. Brit. Mus., **3**:588). – Terra typica: Accra (Type, ♂, in British Museum).

Diagnosis: Head orange; spot on the second palpal joint, third palpal joint, sides of frons, antennae black; collar orange with black stripes; thorax pale yellow; tegulae orange; pectus and legs orange; fore and mid-tibiae black; abdomen orange, with black edges to the rings. Fore wings red; basal area yellow, with two black lines; dorsal area yellow to near tornus; a large yellow spot edged with black before the discocellulars, and another below it; outer area with two large and two or three small yellow spots; silvery lines between the yellow markings; cilia blackish, with a pale median line. Hind wings orange, with a black discocellular mark, and a black marginal band with the inner edge irregular. Rather variable, with more or less black, and the basal and medial yellow areas connected costally. Length of fore wing 15–20 mm.

Distribution: The greater part of Central and Southern Africa, from Senegambia and Erithrea to the Cape Province and Natal.

Genus **Pseudospiris** BUTLER

(fig. 28)

1895 *Pseudospiris* BUTLER, Proc. zool. Soc. London, **1895**:267. – Species typica: *Pseudospiris paidiformis* BUTLER (by monotypy).

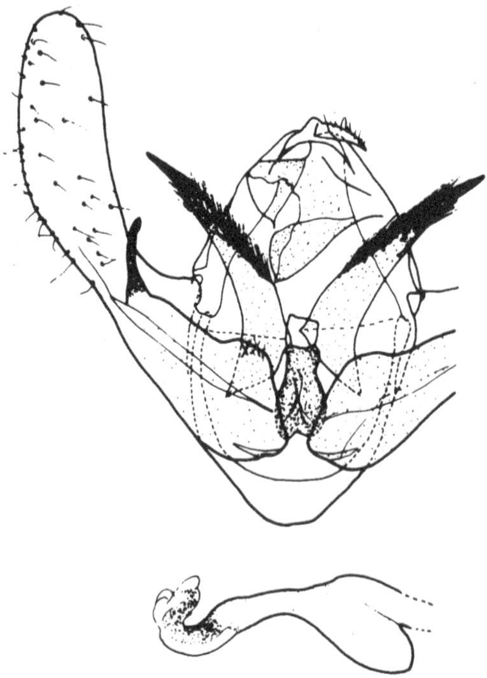

Fig. 28. *Pseudospiris paidiformis* (BUTLER). B. M. Noct. 8462. Typus

D i a g n o s i s : Proboscis full developed. Palpi short, porrect, clothed with rough scales, not reaching as far as frons, which has a truncate conical prominence excised in front, with a flat corneous plate below it with rounded outer edge. Antennae short, simple, faintly dilated towards the extremity. Fore tibiae with a curved spine at the extremity of the outer side; first joint of hind tibiae much shorter than the first joint of fore tibiae. Abdomen with slight lateral tufts of hair on terminal urites; anal tuft large. Nervation: fore wings narrow, produced at the apex; vein 2 issued at about two-thirds of cell; 3, 4, 5 separated, equidistant; discocellulars incurved; areole very small, sometimes absent; 6 from lower cell-angle; 7 and the short stalk of 8 + 9 from the extremity of areole; 10 from about middle of the upper edge of areole. In hind wings, inner area clothed with long hair, in male being extremely long; vein 2 from three-fifths of cell; 3, 4 and 6, 7 from the cell-angles; discocellulars incurved; 5 obsolescent, from middle of the discocellulars; 8 free at base, then anastomosing with the cell.

Male genitalia: Uncus exceptionally small, smaller than the harpe, shaped like a knife-blade. Tubus analis short, very broad, broadly sclerified both dorsally and ventrally. Folds of vincula moderately broad, without stiff hair. Valva rather narrow, especially basally, with the apex rounded; pilosity very sparse; fold of sacculus very broad, reaching to slightly before middle of valva, and bearing terminally a small, slender harpe directed costad. Aedeagus about one-half of valva in length, slender, except in the proximal third, strongly arched terminally, with the curved portion covered with minute wartlike cornuti, giving a shagreened appearance; fultura inferior narrowing distally, with the proximal margin rounded and faintly excised at middle; distal margin cut straight; fultura superior forming a pair of divergent labides, broad basally, tapering to a point terminally, nearly as long as the aedeagus, and partially fringed with stiff hair. Saccus short, rounded.

D i s t r i b u t i o n : Central and East Africa.

Pseudospiris paidiformis BUTLER

1895 *Pseudospiris paidiformis* BUTLER, Proc. zool. Soc. London, **1895**:267; pl. 15, fig. 8 and 9. – T e r r a t y p i c a : British Central Africa, Fwambo (Types, ♂, ♀, in British Museum).

D i a g n o s i s : Head and thorax ochreous; extremities of palpi, spots on head and thorax, and edges of collar and tegulae black; legs ochreous, banded with black; abdomen orange yellow, with black segmental edges. Fore wings ochreous; costa streaked with black; markings black; subbasal line oblique; base of cell with a dot; inner line broadly undulate; in cell and below cell, two ocellate spots; discocellular mark fabiform, with a black central line; outer line outcurved in costal half, with a tooth reaching the discocellular mark, and strongly oblique and incurved from there to two-thirds of dorsum; terminal area with black nervural streaks; Indian red spots in basal area below costa and cell; an Indian red band distally of the outer line, following it in shape; dorsal margin with an Indian red streak; a few silvery scales in the basal half of the wing; cilia black at bases, then ochreous. Hind wings in male Indian red; a rather obsolete discocellular mark; outer margin black, broader between the apex and the lower cell-angle, spotted with reddish; in female, orange, becoming reddish in the anal area; discocellular spot distinct, large; terminal band broad, with orange spots between the veins. Length of fore wing 17–24 mm.

D i s t r i b u t i o n : Zambia, Tanzania, Malawi, Zaire (Shaba).

Genus **Choeropais** JORDAN

(fig. 29)

1913 *Choeropais* JORDAN in SEITZ, Großschmetterl. Erde, **15**:13. – S p e c i e s t y p i c a : *Pseudospiris jucunda* JORDAN (by original designation).

D i a g n o s i s : Proboscis fully developed. Palpi short, porrect, slightly fringed with hair below. Antennae short, faintly dilated distally, blunt at the extremity. Frons snout-shaped, obliquely truncated, with a raised rim et the extremity, about as broad in male as in female. Tibiae nearly smooth-scaled except the hind tibiae which are clothed with rough hair; fore tibiae without apical spine, bit with a short curved claw at the extremity on the inner side. Abdomen without crests. Nervation: in fore wings, vein 2 arising from five-sevenths of cell; 3, 4, 5 well separated, equidistant; discocellulars incurved; vein 6 from slightly below the lower cell-angle; areole narrow; 7 and stalk of 8 + 9 from the extremity of the areole; 10 from the upper margin of the areole, near its extremity. In hind wings, vein 2 issuing from three-fifths of the cell; 3, 4 and 6, 7 from the cell-angles; discocellulars incurved; 5 obsolescent, from middle of the discocellulars; 8 briefly anastomosing with cell near the base.

Male genitalia: Uncus about half the costa in length, slender, pointed, nearly straight. Tubus analis very short and slender. Vincula very narrow, fused only apically, with narrow folds; latter bearing at middle a few hairs, but not true bunches of stiff hair. Valva rather broad, terminally lobe-shaped; base of costa bearing a small lobe ending in a few spinules and fringed with stiff hair; sacculus ending in a slender process; fold of sacculus half as broad as the valva, reaching to middle thereof, and ending in a robust, roebuck's horn-shaped harpe. Aedeagus somewhat shorter than the costa, moderately robust, slightly arched; terminal portion bearing a short and a very long, slender process, and a ventral patch of small cornuti; fultura inferior well developed, elongate, somewhat wider basally; fultura superior forming a strong median process reaching to nearly the apex of the vincula. Saccus hardly differentiated, broadly rounded.

Distribution: Angola.

Choeropais jucunda (JORDAN)

1904 *Pseudospiris jucunda* JORDAN, Novit. Zool., **11**:444. – Terra typica: Angola (Types, ♂, ♀, in British Museum).
1913 *Choeropais jucunda* – JORDAN in SEITZ, Großschmetterl. Erde, **15**:13; pl. 3f.

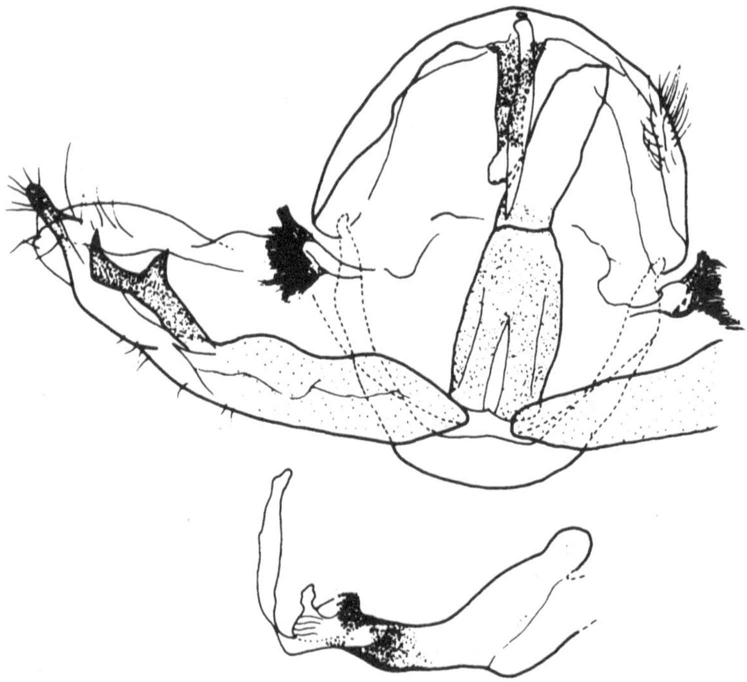

Fig. 29. *Choeropais jucunda* JORDAN. B. M. Noct. 8464

D i a g n o s i s : Head and thorax black; palpi, sides of frons, occiput, collar, tegulae and thorax above marked with yellowish white; legs with whitish streaks; tarsi ringed with whitish; abdomen orange with segmental black lines connected by a dorsal streak except on the basal segments; underside brown with whitish bands. Fore wings pale yellow; costal streak black with a reddish medial area sprinkled with silver; distal two thirds of dorsum reddish; two fine subbasal black lines from costa to cell; a triangular red patch below cell to the antemedial black line, edged proximally with silvery white; medial area with silvery white markings edged with black, viz., three spots on the media, connected with an elliptical cell-spot and another below cell joined to the antemedial line by a streak on the submedian fold, and a rather narrow discocellular lunule; veins beyond cell and vein 1 streaked with black; postmedial line black edged with silvery white distally; some red and silvery sprinkling beyond it; veins in terminal area streaked with red with some black scales towards the termen; cilia black, tipped with yellow. Hind wings red in male, orange in female; termen with eight or nine yellow terminal dots, included in female in the black terminal area; cilia as in fore wings. Length of fore wing 25 mm.

D i s t r i b u t i o n : Angola (Bihé); Zaire (Shaba).

Genus **Massaga** WALKER

(fig. 30)

1854 *Massaga* WALKER, List Lepidopt. Het. Brit. Mus., **2**:358. – S p e c i e s t y p i c a : *Massaga hesparia* WALKER, not *Geometra hesperia* CRAMER = *Massaga maritona* BUTLER (by monotypy)*.

D i a g n o s i s : Proboscis fully developed. Antennae distinctly dilated towards the extremity. Palpi upturned, with the second joint fringed with rough hair, and the third somewhat porrect, short in ♂ and long in ♀. Frons with a conical truncate prominence with raised rim at the extremity. Legs and body covered with appressed hair and scales. Abdomen in ♂ with protrusible lateral tufts of long yellow hair from the base. Nervation: in fore wings, vein 2 arising at three-quarters of cell; 3, 4, 5 well separated, nearly equidistant, or 5 slightly more apart; lower discocellular straight, upper discocellular incurved; 6 from just below the upper cell-angle; areole very small and narrow; 7 and the common stalk of 10, 8 + 9 from the extremity of the areole. In hind wings, vein 2 arising slightly beyond two thirds of cell; 3, 4 and 6, 7 from the cell-angles; discocellulars very faintly incurved; cell clothed with long hair; 5 obsolescent, from middle of the discocellulars.

Male genitalia: Uncus slightly arched, relatively broad basally, tapering to a point terminally, with a slight subterminal expansion, and with long and abundant pilosity. Tubus analis broad. Vincula widening in the medial portion. Valva elongate and narrow, upturned and pointed terminally; costa and sacculus nearly parallel from near the base to the upcurved subterminal portion; sacculus edged with numerous slender spinules in the upcurved portion; harpe hooked, short, occupying the distal

---

* WALKER obviously mistook his specimens for CRAMER's species, and misspelt the name of the latter "*hesparia*" instead of "*hesperia*". WALKER's version must accordingly be considered as synonym of *G. hesperia*.

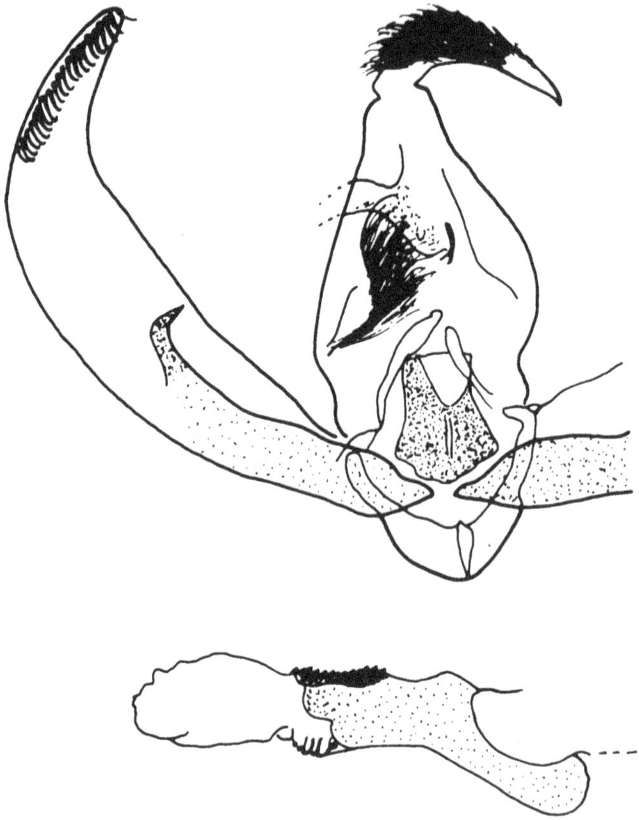

Fig. 30. *Massaga maritona* BUTLER. B. M. Noct. 8604. Typus

half of the fold of the sacculus. Aedeagus about two-thirds of costa in length, robust, with the proximal half spoon-shaped; a terminal ventral patch of minutest cornuti; fultura inferior shaped as a parallelogram with a narrow medial incision and with the distal margin broadly incurved. Saccus short, rounded.

D i s t r i b u t i o n : West and Central Africa to Uganda.

Key to the species

1 Wings without white bands, but with the veins metallic green in the distal portion . . . . ***M. monteirona***, p. 101
1′ Wings with a white or yellowish band 2
2 Band yellowish, narrow, at most 3 mm wide on hind wings; discocellular outside the band in both wings . . . . ***M. angustifascia***, p. 101
2′ Band pure white, wider than 3 mm on hind wings; discocellulars in band in both wings . ***M. maritona***, p. 101

Massaga maritona BUTLER

1854 *Massaga hesparia* WALKER (non *Geometra hesperia* CRAMER), List Lepidopt. Het. Brit. Mus., 2:358. – Terra typica: West Africa (Type, ♂, in British Museum).
1868 *Massaga maritona* BUTLER, Proc. zool. Soc. London, **1868**:17; fig. 1. – Terra typica: Sierra Leone (Type, ♂, in British Museum).
1874 *Massaga virescens* BUTLER, Lepidopt. exot.:175; pl. 61, fig. 15. – Terra typica: not designated, = Cape Coast, fide KIRBY 1892 (Type, ♂, in British Museum).

Diagnosis: Body and wings black with blue or green reflections; head, except the frontal prominence, collar, palpi, except the last joint, fore coxae and femora carmine-red; extremity of abdomen and base of anal tuft yellow; in ♀ the whole anal tuft yellow. Fore wings with the veins streaked with metallic blue; a medial white or yellowish white band, including the discocellulars, narrowing towards the costa but not reaching it, about $2\,^1/_2$ mm wide at middle in ♂, slightly wider in ♀. Hind wings with an about 5 mm wide in ♂, and 6 mm in ♀ white band, including the discocellulars, sometimes narrowing towards the anal angle and ending near it. Length of fore wing 19–28 mm.

Distribution: Sierra Leone, Gold Coast, Nigeria, Cameroon, North and Central Zaire, Uganda.

Massaga angustifascia ROTHSCHILD

1896 *Massaga angustifascia* ROTHSCHILD, Novit. Zool., 3:31. – Terra typica: Old Calabar (Type, ♀, in British Museum).

Diagnosis: Body and wings black with bronze-green reflections; head, collar, pectus in front, fore coxae and femora carmine-red. Fore wings with a pale yellow band about 3 mm wide, placed 1–2 mm beyond the discocellulars, bent inwards and narrowing towards costa but not reaching it. Hind wings with a pale yellow band mot including the discocellulars, slightly angled at vein 5 and ending near the anal angle. Length of fore wing 28 mm.

Distribution: Old Calabar.

Massaga monteirona BUTLER

1874 *Massaga monteirona* BUTLER, Cist. Entomol., 1:216. – Terra typica: Ambriz, fide KIRBY 1892 (Type, ♂, in British Museum).
1878 *Eusemia metallica* MABILLE, Bull. Soc. Zool. France, Paris, 3:88. – Terra typica: Kabinda, Congo = Zaire (Type, ♂, in British Museum).
1892 *Massaga monteironis* – KIRBY (ex errore pro *monteirona*), Syn. Cat. Lepidopt. Het.:33.
1892 *Massaga metallica* – KIRBY, Syn. Cat. Lepidopt. Het.:33.

Diagnosis: Body and wings black with blue reflections; head, except the terminal palpal joint and the frontal prominence, collar, pectus in front, fore coxae and femora carmine-red; abdomen with a yellow band on the underside of the last urite. Fore wings with the veins streaked with metallic greenish-blue; a diffused obsolescent metallic greenish band just beyons the cell. Hind wings with stronger blue reflections, with a postmedial series of short streaks on veins. Length of fore wing 28 mm.

D i s t r i b u t i o n : Angola to Zaire, Cameroon and Ivory Coast.

Genus **Massagidia** HAMPSON

(fig. 31)

1901 *Massagidia* HAMPSON, Cat. Lepidopt. Phal. Brit. Mus., 3:644. – S p e c i e s t y p i c a : *Geometra hesperia* CRAMER (by original designation).

D i a g n o s i s : Proboscis fully developed. Antennae distinctly dilated towards the extremity. Palpi with the second joint obliquely upturned, fringed with hair, the third porrect, short in ♂, long in ♀. Frons with a truncate conical projection with a raised rim at the extremity. Hind tibiae fringed with hair above. Abdomen smoothly scaled. Nervation: In fore wings, vein 2 given off at seven-tenths of cell; 3, 4, 5 rather well separated, 4–5 slightly wider apart than 3–4; lower discocellular straight; areole short and narrow, sometimes absent; 6 from the upper cell-angle; 7 and the stalk of 8 + 9 from the extremity of the areole; 10 from the upper angle of the same. In hind wings, vein 2 arising at five-sevenths of cell; 3, 4 and 6, 7 from the cell-angles; discocellulars rather faintly incurved; 5 obsolescent from middle of the same.

Male genitalia: Uncus about two-fifths of costa in length, or slightly longer, rather broad basally, but narrowing towards the extremity which is arched and pointed, pilosity long and abundant. Tubus analis broad, well sclerified. Fold of vinculum moderate, with a small hair tuft. Valva banana-shaped, rather distinctly uprcurved terminally; sacculus edged for about one-fifth in the terminal portion with long spine-shaped hair; fold of sacculus rather broad at the base, then narrowing, not reaching to middle of the valva, and ending in a small, slender, arched harpe. Aedeagus about one-half of costa in length, rather robust, spoon-shaped in the proximal half; a lateral patch of cornuti at the extremity; fultura inferior with a rounded proximal piece, otherwise shaped like a parallelogram with the distal margin excised; fultura superior sclerified at edges, forming a kind of crescent. Saccus rather short, semi-circular. *M. xenia* shows a differently-shaped uncus, with the basal half bearing a broad dorsal ridge, and the terminal half tapering to a point. The whole shape of the uncus reminds of the bill of a calao, with the basal casque. Otherwise similar to the type, but the patch of cornuti on the terminal portion of the aedeagus is much reduced, and the fultura inferior has the proximal margin rounded, and the distal margin straight.

D i s t r i b u t i o n : West and Central Africa.

Key to the species

1 White band on fore wings distinctly curved (♀) or forming an angle (♂) 2
1' White band on fore wings oblique, straight . . . . *M. xenia*, p. 104
2 White band on hind wings not narrowing towards anal margin . . . . . . *M. hesperia*, p. 103
2' White band on hind wings narrowing and pointed towards anal margin *M. tenuifascia*, p. 104

Massagidia hesperia (CRAMER)

1775 *Geometra hesperia* CRAMER, Uitlandsche Kapellen, **1**:pl. 56 C. – Terra typica: West Africa (Type not designated).
1868 *Massaga delicia* BUTLER, Proc. zool. Soc. London, **1868**:224; pl. 17, fig. 2. – Terra typica: Sierra Leone = Gold Coast, fide KIRBY 1892 (Type, ♂, in British Museum).
1892 *Massaga hesperia* – KIRBY, Syn. Cat. Lepidopt. Het.:33.
1894 *Massaga demena* DRUCE, Ann. Mag. nat. Hist., London, (6) **14**:23. – Terra typica: Cameroon (Types, ♂, ♀, in British Museum).
1901 *Massagidia hesparia* – HAMPSON, Cat. Lepidopt. Phal. Brit. Mus., **3**:644; fig. 284.

Diagnosis: Body black; head, except the third palpal joint, antennae and frontal prominence, collar, fore coxae and femora carmine-red; abdomen with a segmental band on the last urite and the anal tuft in ♂, the whole extremity in ♀, carmine-red. Wings with brilliant blue reflections; fore wings with veins streaked

Fig. 31. *Massagidia hesperia* (CRAMER). B. M. Noct. 8474

with metallic blue; a medial white band not quite reaching costa in ♂, angles outwards distally beyond the lower angle in cell, in ♀ bent inwards from vein 4 to costa; cilia white apically. Hind wings with a white medial band rounded above and not reaching costa, its outer edge excurved beyond cell; the band is broader and extending to the anal angle in ♀. Length of fore wing 16–22 mm in ♂, 21--27 mm in ♀.

Distribution: West Africa to Zaire, Uganda and Burundi.

Massagidia tenuifascia HAMPSON

1910 *Massagidia tenuifascia* HAMPSON, Proc. zool. Soc. London, **1910**:401; pl. 36, fig. 12. – Terra typica: Kambove District, Congo = Zaire (Type, ♀, in British Museum).

Diagnosis: Head and collar carmine-red; antennae and the last palpal joint black; thorax and abdomen black, slightly shot with blue; pectus in front and fore femora carmine-red; terminal urites carmine-red. Fore wings black with blue reflections; veins and the submedial fold streaked with brilliant blue; a rather broad medial white band, oblique from costa to the lower angle of the cell, then erect; cilia white apically. Hind wings black with blue reflections; a medial white band narrowing to the anal margin and with the distal edge curved; cilia white apically. Length of fore wing 19–25 mm.

Distribution: Southern Zaire (Lulua). Only ♀♀ known.

Massagidia xenia (JORDAN)

1913 *Massaga xenia* JORDAN in SEITZ, Großschmetterl. Erde, **15**:22. – Terra typica: Ja River, Bitye, Cameroon (Type, ♂, in British Museum).
1920 *Massagidia xenia* HAMPSON, Cat. Lepidopt. Phal. Brit. Mus., Suppl. **2**:590; pl. 71, fig. 11.

Diagnosis: Head, collar and prothorax carmine-red; rest of thorax black with metallic blue reflections; antennae black; abdomen black with metallic blue reflections; anal tuft carmine-red. Wings black with violet reflections. In fore wings, veins and dorsum streaked with metallic blue; a yellowish-white band, broader in ♀, edged proximally with diffused metallic blue, from middle of costa to dorsum beyond the middle, expanding distally in interspace III; cilia white at extremities apically. Hind wings with a large tuft of hair from the anal margin before the anal angle in ♂; veins in the terminal area streaked with blue; a broad white in ♂, yellowish-white in ♀ band from middle of costa to terminal point of the anal margin; tuft of hair on the anal margin black; cilia white terminally at apex. Length of fore wing 23–25 mm.

Distribution: Cameroon; Kasai Province, Zaire.

Genus Paratuerta HAMPSON

(fig. 32)

1902 *Paratuerta* HAMPSON, Ann. South Afr. Mus., Capetown, **2**:282. – Species typica: *Paratuerta marshalli* HAMPSON (by original designation).

Diagnosis: Proboscis fully developed; antennae of ♂ more or less strongly dentate, sometimes nearly filiform, in ♀ minutely serrate; palpi porrect, with the second joint fringed with long scales below, and the third moderate in ♂ and long in ♀; frons with a large truncate conical prominence, with a raised rim at the extremity, and a corneous plate below it; thorax clothed with rough hair and scales mixed, without crests; tibiae fringed with long hair; fore tarsi clothed with rough hair above; abdomen with dorsal crests on the basal urites. Nervation: in fore wings, vein 2 given off at about five-sixths of cell; 3, 4, 5 separate, equidistant; 6 from the upper cell-angle; areole very narrow; 7 and the short stalk of 8 + 9 from its extremity; 10 from its upper margin. In hind wings, vein 2 from about five-sevenths of cell; 3, 4 and 6, 7 stalked, or 3, 4 slightly separated; discocellulars incurved; 5 obsolescent, from middle thereof; 8 anastomosing with the cell at a point near the base.

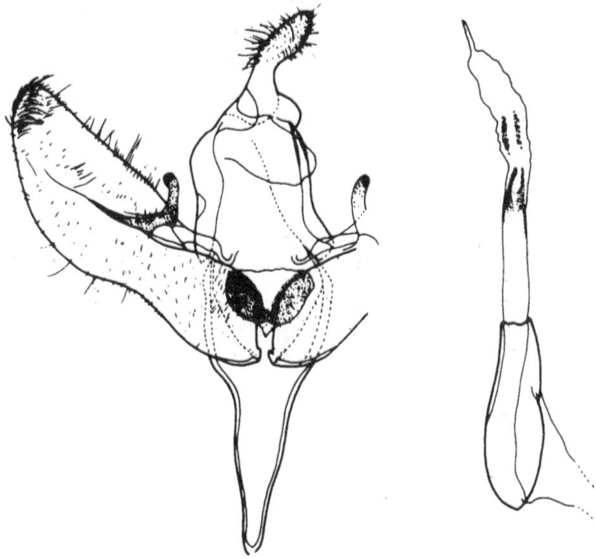

Fig. 32. *Paratuerta marshalli* HAMPSON. B. M. Noct. 8368

Male genitalia: Uncus short, narrow at the base, then compressed, with a small terminal hook and long pilosity. Tubus analis short and broad. Vincula rather narrow, with a narrow fold. Valva elongate and rather broad; costa nearly straight or faintly outcurved at one-third, and faintly incurved elsewhere; apex rounded, bearing a fringe of rather stiff short hair; sacculus outcurved for one-half, then incurved, and again outcurved; fold very broad, nearly reaching the costa at the base, and extending to nearly four-fifths of the sacculus, bearing at middle a short, blunt, faintly arched harpe. Aedeagus longer than the valva (about 5/4), very slender, nearly straight, with the terminal portion bearing two narrow patches of minute wartlike cornuti; fultura inferior broadly hear-shaped, with the distal margin nearly straight. Saccus much elongate, narrowly triangular, as long as one-half of the valva.

Distribution: East Africa.

Key to the species

1 Fore wings with a white stripe, not bent upwards distally; hind wings orange yellow with a narrow black margin . . . . *P. marshalli*, p. 106

1′ Fore wings with the white stripe bent upwards distally; hind wings yellowish with only a reddish brown spot on the anal angle . . . . *P. featheri*, p. 106

Paratuerta marshalli HAMPSON

1902 *Paratuerta marshalli* HAMPSON, Ann. South Afr. Mus., Capetown, 2:282. – Terra typica: Salisbury, (Types, ♂, ♀, in British Museum).

Diagnosis: Antennae serrate. Head black mixed with white; antennae black; thorax dark brown mixed ochreous white; pectus and legs orange; fore femora fringed with purple-brown; fore tibiae and the tarsi black sprinkled with white; abdomen orange, with dorsal and ventral series of black spots. Fore wings purplish grey sprinkled with white; cell and area just below and beyond it to the postmedial line tinged with olive-brown, with slight dark streaks on the veins and in the submedian fold; an irregular white fascia in the submedian interspace from base to the postmedial line, dorsally produced in an oblique tooth at middle, then excised; extremity of fascia acute with two small white spots above it on the postmedial line, which is indistinctly double, more or less parallel with the termen, not extending below the white streak; subterminal line indistinct, blackish, minutely waved; towards the tornus, a blackish terminal line; cilia dark brown mixed grey. Hind wings orange yellow, with a narrow black terminal band from the apex to vein 1, the terminal part of which latter sprinkled with black scales. Length of fore wing 20–24 mm.

Distribution: Rhodesia, Mashonaland, Kenya.

Paratuerta featheri FAWCETT

1915 *Paratuerta featheri* FAWCETT, Proc. zool. Soc. London, **1915**:94; pl. 1, fig. 7. – Terra typica: Kenya (Type, ♂, in British Museum).

Diagnosis: Antennae filiform. Head and thorax red-brown; abdomen yellow, with two black dorsal spots on the second and the third urites, and four fuscous dorsal bands on the succeding urites. Fore wings red-brown, tinged with pink; a pale pinkish costal subapical fascia; terminal area pale pinkish, with a red-brown tornal spot; a broad silvery-white fascia from base of costa along vein 1, reaching to two-thirds of dorsum, then bent up towards the apex between veins 4 and 5, and bordered above and distally by a dark sap-green fascia, which also bends up and reaches the costa before the apex. Hind wings dull yellow, with a red-brown anal spot, surrounded by a small patch of red-brown sprinkling. Length of fore wing 21 mm.

Distribution: Kenya.

Genus Hemituerta gen. nov.

(fig. 33)

Species typica: *Paratuerta nana* HAMPSON (designatione hoc loco).

Diagnosis: Proboscis fully developed; antennae filiform, pointed; palpi densely scaled, about as long as twice the diameter of eye; the first urite with a raised scaly tuft. Nervation: in fore wings, vein 2 arising at about four-fifths of cell; 3, 4, 5 separate, 4, 5 slightly more distant from each other than 3, 4; 6 from the lower cell-angle; areole short; 7 and the short stalk of 8 + 9 from its extremity; 10 from its upper angle. In hind wings, vein 2 arising at about five-sixths of cell; 3, 4 much approximated; discocellulars angled inwards; 5 from their middle; 6, 7 from a point; 8 anastomosing with cell at a point near the base.

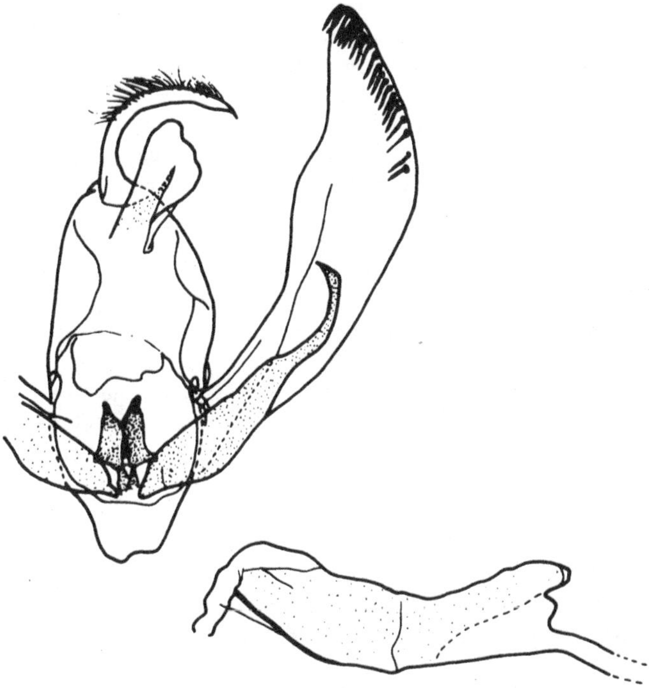

Fig. 33. *Hemituerta nana* (HAMPSON). B. M. Noct. 8558. Typus

Male genitalia: Uncus moderately long, about seven-elevenths of costa in length, slender, compressed, regularly arched, with a small terminal hook. Tubus analis short and broad, especially terminally, partly sclerified. Fold of vinculum narrow. Valva very elongate and narrow, slightly broadening subterminally; costa outcurved in that portion, otherwise moderately incurved; apex pointed; sacculus more or less regularly outcurved, bearing in the last quarter to apex a fringe of strong spines; fold broad, reaching slightly beyond one-third of the valva, and ending in a well-developed, sickle-shaped harpe. Aedeagus about as long and sometimes distinctly shorter than costa, robust, spoon-shaped proximally, faintly arched; fultura inferior rather small, with the proximal margin slightly notched in middle, and with the angles produced; distal margin with a triangular median incision; sometimes the fultura inferior is horseshoe-shaped, with a distal convexity. Saccus short, angulate.

D i s t r i b u t i o n : Kenya, Somali.

Key to the species

1 Abdomen white; hind wings ochreous-yellow with a red-brown subterminal band . . . . *H. mahdi*, p. 108

1′ Abdomen yellow; hind wings orange-yellow with a blackish-brown terminal band . . . . *H. nana*, p. 108

Hemituerta nana (POULTON)

1916 *Paratuerta nana* POULTON, Prox. zool. Soc. London, **1916**:112; pl. 1, fig. 17. – Terra typica: Somaliland (Type, ♂, in British Museum).

Diagnosis: Head and thorax white, mixed with some brown scales; abdomen white, sprinkled dorsally with brown; the double basal crest with some metallic-blackish scales. Fore wings white sprinkled with brown, more thickly in the terminal area; costal area and disc in ♂ grey sprinkled with brown; a sinuous blackish-brown streak in the submedian fold to the postmedial line, with the area below it, and also area from just before the postmedial line to the subterminal line, chocolate-brown mixed with grey; antemedial line obsolete to the submedian fold, then blackish and strongly angled outwards above dorsum; cell- and discocellular spots in ♂ defined by dark brown, the latter faintly distally, the former round; in ♀, spots faint, diffused, reddish-brown, almost conjoined; postmedial line blackish-brown, obliquely curved, slightly waved, from costa to the streak in the submedian fold, before which the brown is angled inwards at the discal fold; subterminal line blackish-brown, obliquely curved, slightly waved, angled inwards on vein 1; a fine dark terminal line. Hind wings brownish-ochreous in ♂, ochreous yellow in ♀, with the anal area tinged with reddish-brown, and the postmedial area reddish-brown to near termen which is yellowish sprinkled with brown; a terminal series of brown striae; cilia on both wings chequered brown and white. Length of fore wing 11–13 mm.

Distribution: Somaliland.

Hemituerta mahdi (PAGENSTECHER)

1903 *Aegocera mahdi* PAGENSTECHER, Jb. nassau. Ver. Naturkde., Wiesbaden, **16**:9 pl. 1, fig. 3. – Terra typica: Somaliland: Hanajo (Type, ♀, in British Museum).

1915 *Paratuerta argentifascia* FAWCETT, Proc. zool. Soc. London, **1915**:94; pl. 1, fig. 7a. – Terra typica: Kenya, Kedai (Type, ♂, in British Museum).

1920 *Tuerta mahdi* – HAMPSON, Cat. Lepidopt. Phal. Brit. Mus., **2**:588.

Diagnosis: Head and thorax red-brown mixed with white; frons white at sides; second and third palpal joints white terminally; pectus white; legs brown and white; fore tibiae banded yellow and blackish; abdomen yellow orange; basal crest white and black, followed by some brown sprinkling. Fore wings chocolate brown thickly sprinkled with grey-white scales; inner medial area redder; a wedge-shaped white fascia in cell from base, then along the media to the lower cell-angle, its outer edge oblique and sinuous, met at the lower cell-angle by an oblique white band from costa towards apex, its outer edge slightly angled at vein 4; an oblique buff antemedila band edged proximally by silver, defining the distal edge of the postmedial band and the fascia, then continued to dorsum; some silvery scales below costa near the base, in upper cell-angle, and on discocellulars; an indistinct red subterminal band, sprinkled with silvery; a series of small black spots just before termen. Hind wings orange-yellow; a blackish-brown terminal band, narrowing to tornus. Length of fore wing 14–15 mm.

Distribution: Southern Ethiopia: Somaliland; Kenya.

Genus **Hypotuerta gen. nov.**

(fig. 34)

Species typica: *Tuerta transiens* HAMPSON (designatione hoc loco).

Diagnosis: Proboscis fully developed; antennae in ♂ reaching to three-quarters of costa, distinctly dilated distally, with the first joint swollen; in ♀ shorter, filiform; eyes very large, globulose, nearly conjoined on vertex; palpi upturned, about as long as the diameter of eye, with pilosity moderate but rough; last joint minute, slender and naked; legs with the pilosity abundant, moderately long on fore and mid legs, longer on hind legs; hind tibiae with very long,

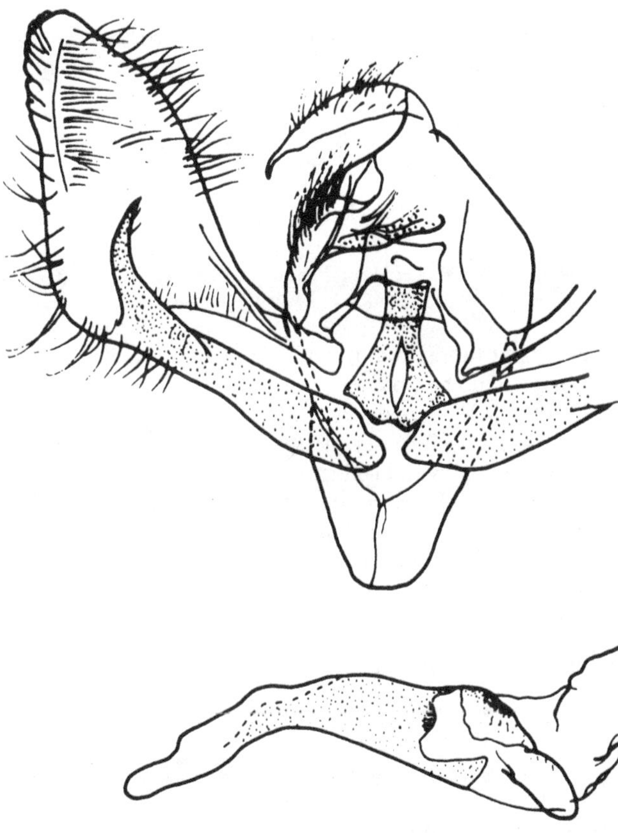

Fig. 34. *Hypotuerta transiens* (HAMPSON). B. M. Noct. 8569

brush-like pilosity; that of hind tarsi somewhat shorter; abdomen with a series of dorsal crests, becoming smaller distally. Fore wings roughly triangular; costa nearly straight, very faintly outcurved; apex rounded; termen more oblique in ♂, less outcurved, sometimes nearly straight; tornus rounded; dorsum hardly outcurved at base. Nervation: in fore wings, vein 2 given off at four-fifths to five-sixths of cell; 3, 4, 5 much approximated, nearly equidistant; discocellulars faintly incurved; 6 from the lower cell-angle; areole long and narrow; 7 and the short stalk of 8 + 9 from its extremity; 10 from its upper margin. In hind wings, vein 2 from very near the cell-angle; 3, 4 very short stalked; discocellulars obsolescent, faintly incurved; 5 obsolescent, from middle thereof; 6, 7 short stalked; 8 fused for a short while with cell at base.

Male genitalia: Uncus short, compressed, arched basally, then nearly straight, with a slender, elongate terminal hook. Tubus analis short and broad, sclerified ventrally. Vinculum dilated in middle portion, with a fringe of stiff hair. Valva rather narrow basally, then expanding and diamond-shaped with the angles rounded; subapical part of sacculus and apex with a fringe of about ten stiff setae; fold rather narrow, with a short, arched harpe. Aedeagus very slightly longer than costa, terminally robust, slightly arched in the proximal portion; fultura inferior narrowing distally; proximal margin notched at middle, and with the angles somewhat produced; terminal margin cut straight. Saccus elongate semi-elliptical.

This new genus differs from the genus *Tuerta* WALKER chiefly in the presence of stiff setae before and on apex of the valva, and in veins 3, 4 and 6, 7 in hind wing being stalked.

Distribution: West, Central and East tropical Africa.

Hypotuerta transiens (HAMPSON)

1901 *Tuerta transiens* HAMPSON, Cat. Lepidopt. Phal. Brit. Mus., 3:624; pl. 53, fig. 15. – Terra typica: Old Calabar (Type, ♀, in British Museum).

Diagnosis: Head and thorax red-brown; palpi fringed with orange hair; pectus and base of legs orange; abdomen orange with a dorsal red-brown stripe; extremity orange in ♂, red-brown in ♀. Fore wings chocolate-brown; a silvery-white fascia from base of costa in and below cell to beyond middle in the submedian interspace, ending in a point; a large lunule beyond cell from interspace II to vein 8; an oblique silvery-leaden spot in middle of cell; a silvery-leaden discocellular bar; a postmedial line incurved below vein 3, and at vein 2 retracted, then excurved again; beyond, a series of dentate spots; a metallic-leaden terminal band. Hind wings orange-yellow; a terminal band, red-brown in ♂, blackish-brown in ♀, narrowing to anal angle. Length of fore wing 20–24 mm.

Distribution: Old Calabar, Nigeria, Cameroon, Gaboon, Zaire: Uele.

Genus Tuerta WALKER

(fig. 35)

1869 *Tuerta* WALKER, Proc. nat. Hist. Soc. Glasgow, 1:346. – Species typica: *Tuerta chrysochlora* WALKER (by monotypy).

D i a g n o s i s : Proboscis fully developed; antennae nearly setiform in both sexes; palpi obliquely upturned, with the first and the second joints fringed with erect hair in front, and the third roughly scaled; frons with a weak truncate conical projection, with a raised rim at the extremity, and not narrowed caudally in ♂; tibiae fringed with hair above; hind tibiae rough in both sexes; abdomen with dorsal crests except on terminal urites. Nervation: in fore wings, vein 2 arising from about three-quarters of cell; 3, 4, 5 approximated, equidistant; discocellulars faintly incurved; 6 from the lower cell-angle; areole narrow; 7 and 8 + 9 short stalked, from the extremity thereof; 10 from the upper angle thereof. In hind wings, vein 2 arising from about five-sevenths of cell; 3, 4 and 6, 7 from the cell-angles; discocellulars incurved; 5 obsolescent, from the middle thereof.

Male genitalia: Uncus small, short, lance-shaped, depressed, with lateral pilosity long and dense. Tubus analis short and very broad, partly sclerified. Vinculum expanding distally to form a squarish lobe at each side, with the distal margin covered with minute spicules. Valva nearly triangular, with the costa about straight, the apex pointed, and the sacculus faintly outcurved; at base of costa, a slender pointed, hairy process reaching to middle of valva; fold of sacculus broad at base, then narrowing to reach beyond two-thirds of valva, then slightly expanded again and bearing a curved harpe covered with minute spicules. Aedeagus about as long as the sacculus, slender, nearly straight, tapering to a point terminally; distal two-thirds covered with semimicroscopic warts; fultura inferior wedge-shaped, with the distal margin broadly excised. Saccus small, narrow, triangular.

D i s t r i b u t i o n : West, Central and East Africa.

Key to the species

1 Fore wings partly olive-green and dark chocolate brown . . . . *T. chrysochlora*, p. 111
1' Fore wings deep chocolate-brown, tinged and partly marked with silvery-blue . . . . . .
T. cyanopasta, p. 112

Tuerta chrysochlora WALKER

1869 *Tuerta chrysochlora* WALKER, Proc. nat. Hist. Soc. Glasgow, 1:347; pl. 6, fig. 7. – T e r r a t y p i c a : "Congo" (Type, ♂, in British Museum).

D i a g n o s i s : Head and thorax blackish-chocolate brown, mixed with a few white scales; pectus and legs orange; fore tibiae and tarsi ochreous and brown; abdomen orange with a dorsal series of black spots. Fore wing olive-green; dorsal area and terminal third dark chocolate-brown; some pure white markings, viz., two cell-points; discocellular lunule; some scales beyond cell; and a short streak on base of vein 4; dark brown dorsal area sprinkled with white, broadened at middle and occupying there the middle portion of interspace Ib; before tornus an oblique carmine-red line, often limited distally by a darker subterminal line running parallel to termen; dark brown terminal area very finely sprinkled with white except proximally; apex purplish-black, with a few fine white streaklets; cilia concolorous. Hind wings orange with a broad dark brown terminal band emitting a tooth on vein 1 to near base; cilia pale grey, darker and browner towards the anal angle. Length of fore wing 15–19 mm.

D i s t r i b u t i o n : Gaboon, Nigeria, Gold Coast, Sierra Leone, Cameroons, Zaire (Kinshasa, Kibali-Ituri, Uele, Tshuapa), Burundi.

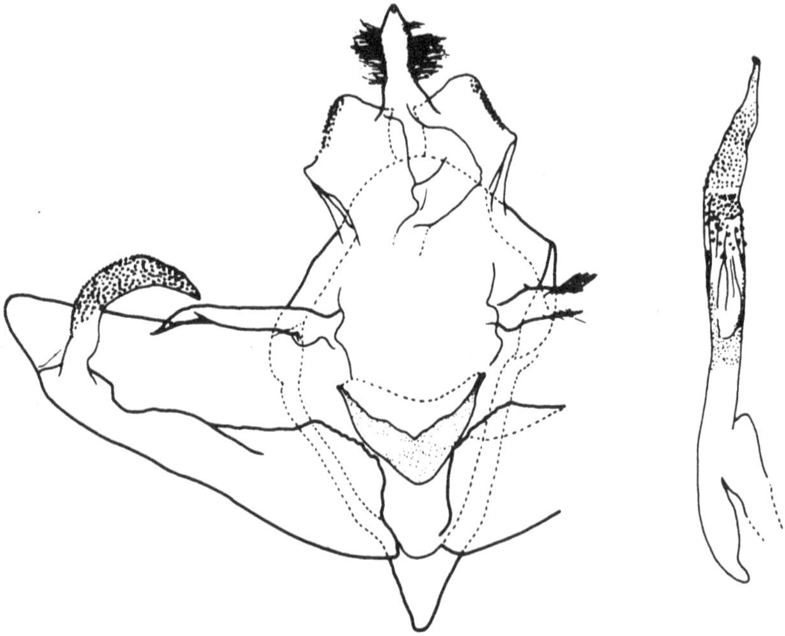

Fig. 35. *Tuerta chrysochlora* WALKER. B. M. Noct. 8288. Typus

Species incertae sedis
Tuerta cyanopasta HAMPSON

1907 *Tuerta cyanopasta* HAMPSON, Ann. Mag. nat. Hist., London, (7) **19**:241. – Terra typica: N'joro, British East Africa = Kenya (Type, ♀, in British Museum).

Diagnosis: Head and thorax blackish-brown; pectus pale orange at sides; fore tibiae and the tarsi ringed greenish; hind tibiae partly orange, and fringed with orange hair; abdomen orange with a series of black spots on dorsum and the extremity blue-black; underside brownish in proximal half, blue-black distally. Fore wings deep chocolate-brown tinged with silvery-blue to the subterminal line; base of dorsum blackish-brown; a waved antemedial line from costa to the media; an almost median waved bar from the submedian fold to dorsum; cell-extremity and discocellulars with blue markings broadly edged with black, the former obliquely elliptical and conjoined at the lower extremity to the latter which is lunulate; postmedial line strong, black, strongly incurved below vein 3; outer edge of the blue area dentate with black dots in the interspaces; a terminal series of black dots; a fine waved terminal line; cilia brownish-blue with pale base. Hind wings orange-yellow with a brownish-black terminal band, narrowing to a point at the anal angle; cilia with pale base. Length of fore wing 21 mm.

Remark: There is only the single type in the collection of the British Museum (Natural History); it being a ♀, the generic appurtenance of the species is uncertain.

Distribution: Kenya.

Genus **Pseudotuerta gen. nov.**
(fig. 36)

Species typica: *Tuerta argyrochlora* CARCASSON (designatione hoc loco).

Diagnosis: Proboscis fully developed; antennae filiform; palpi porrect, as long as the diameter of the eye; first and second joints densely pilose; third joint short, knoblike; pilosity of legs short; abdomen with a small basal crest. Nervation: in fore wings, vein 2 arising at four-fifths of cell; 3, 4, 5 well separated; 5 slightly nearer 4 than 4 to 3; discocellulars faintly angled inwards; 6 from the upper cell-angle; areole narrow; 7 and the stalk of 8 + 9 short stalked, from the extremity of the areole; 10 from the upper margin thereof, near its extremity. In hind wings, vein 2 arising at five-sixths of cell; 3, 4 very slightly separated; discocellulars faintly angles inwards; 6, 7 from a point; 8 anastomosing with cell at a point, soon after the base.

Male genitalia: Uncus rather short and broad, arched at base, with a minute terminal hook. Tubus analis about as long as the uncus, partly sclerified. Vinculum rather narrow, with short pilosity on the broader median area. Valva relatively very large, moderately broad in basal third, then expanding, with the costa distinctly outcurved, the termen broadly rounded, and the sacculus faintly angled inwards at middle; fold of sacculus broad to middle where it becomes much narrower and continues to nearly reach the termen, and bears subterminally a long and strong, faintly arched harpe – quite an unusual position. Aedeagus somewhat longer than uncus (about 4/3), moderately robust, faintly arched, swollen terminally, with the swollen portion covered with short spinelike cornuti; fultura inferior hardly differentiated. Saccus semi-oval.

Remark: *Tuerta argyrochlora* CARCASSON is here placed in a distinct genus because of the very aberrant male genitalic structures.

Distribution: East and Central Africa.

Pseudotuerta argyrochlora (CARCASSON)

1964 *Tuerta argyrochlora* CARCASSON, J. east afr. nat. Hist. Soc., **24** (5) (109):72–73; figs. 30 and 59. – Terra typica: Uganda, Kalinzu Forest, Ankole (Type, ♂, in British Museum).

Diagnosis: Palpi and head blackish-purple-brown, slightly sprinkled with white; base of the antennae with a white tuft; above the eye another white tuft; thorax as head, with a few white hairs and scales; pectus and legs deep chrome-yellow; tibiae and tarsi brownish, speckled with dark purplish-brown; abdomen orange yellow with a dorsal series of dark brown spots; anal tuft orange yellow; underside unspotted. Fore wings parrot-green, with the costa paler; a large irregular discocellular spot; two small white dots in cell, another on the origin of vein 2; a small white streak in interspace 1 b; basal portion of dorsum very dark purple-brown; a large dark purple-brown spot with darker edges, and speckled with white, in the central part of interspace 1 a, expanding into the lower half of interspace 1 b; a very dark brown, almost black, irregular curved line from costa at three-quarters from base to dorsum at two-thirds from base; marginal area distal to the dark line

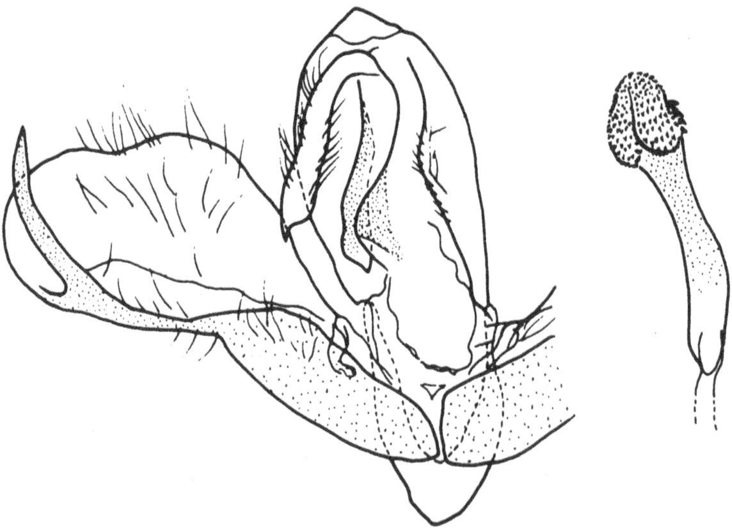

Fig. 36. *Pseudotuerta argyrochlora* (CARCASSON). B. M. Noct. 4526. Typus

dark purple-brown; a faint wavy, very dark purple-brown submarginal band, edged distally with a few white scales, especially near the apex; cilia purple-brown. Hind wings orange yellow, with a broad blackish-brown terminal band, from apex to tornus, narrowing towards the latter; a faint blackish streak from tornus to the middle of anal margin; cilia pale grey brown; long, dense hairs, especially near the anal margin. Length of fore wing 16–20 mm.

Distribution: Uganda; Zaire (Kibali-Ituri, Kivu)

Genus **Tuertella gen. nov.**

(fig. 37)

Species typica: *Tuerta rema* DRUCE (designatione hoc loco).

Diagnosis: Antennae filiform; palpi upturned, about as long as the diameter of the eye, thick, densely hairy; last joint minute, naked, pointed; hind tibiae fringed with longish hari; abdomen with a basal crest. Nervation: in fore wings, vein 2 given off at two-sevenths of cell; 3, 4 widely separated; 4, 5 from about a point; discocellulars faintly incurved; 6 from the lower cell-angle; areole elongate rhomboidal; 7 and the short stalk of 8 + 9 from its extremity; 10 from its upper margin, slightly beyond the middle. In hind wings, vein 2 given off at four-fifths of cell; 3, 4 and 6, 7 from the cell-angles; discocellulars incurved; 5 obsolescent, from their middle; 8 anastomosing with cell at a point, shortly after the base.

Male genitalia: Uncus relatively short, depressed, nearly straight, with a minute terminal hook. Tubus analis short, very broad, narrowing distally, sclerified ventrally. Vinculum narrow, without well marked pilosity. Valva elongate and broad, especially distally; costa very faintly incurved, bearing at base a hairy fold; apex

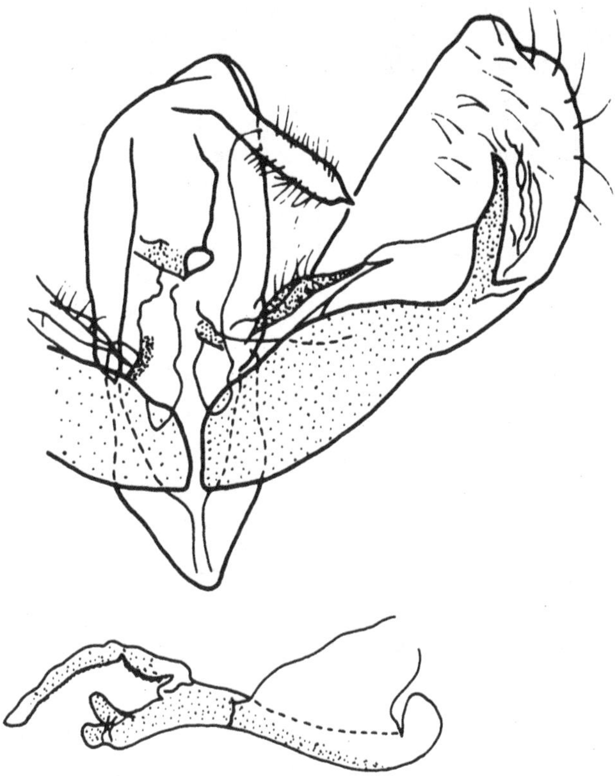

Fig. 37. *Tuertella rema* (DRUCE). B. M. Noct. 8582. Typus

rounded and slightly produced; sacculus faintly outcurved in the proximal half, much more distinctly distally; fold very broad, slightly narrowing distally, reaching middle of valva, and bearing terminally a faintly arched, somewhat irregularly shaped harpe. Aedeagus slightly shorter than the costa, slender, faintly S-shaped, with the terminal margin slightly bifid, and with a slender subterminal process, minutely toothed on the near side; fultura inferior consisting of two lateral oval plates. Saccus more or less triangular.

Tuerta rema DRUCE is here being placed in a new genus because of the rather different male genitalic structures.

Distribution: East Africa.

Tuertella rema (DRUCE)

1910 *Tuerta rema* DRUCE, Ann. Mag. nat. Hist., London, (8) **6**:169. – Terra typica: German East Africa = Tanzania (Type, ♂, in British Museum).

Diagnosis: Head black, sprinkled with white; thorax rusty brown, sprinkled with white; pectus, legs and abdomen pale yellow. Fore wings with the costal area white to beyond middle and below the cell, edged by sap-green; terminal area also white; dorsal and postmedial areas purple, the former sprinkled with white; antemedial line double, chocolate-red, filled in with green in dorsal area; some pale greenish tinge in and below cell-extremity, with a white point on the media and the base of vein 2; the green colour on the outer edge of the white area emitting a streak on dorsal side to the lower cell-angle; postmedial line chocolate-red, with a black patch sprinkled with white beyond it on the costa; subterminal line slight, white, waved, edged distally by chocolate-red lunules. Hind wings pale yellow with a slight brown streak on the extremity of vein 1 and a brown subterminal bar above vein 1 with a little red beyond it on the termen; cilia white. Length of fore wing 18 mm.

Distribution: Tanzania; Burundi.

Genus **Charitosemia** KIRIAKOFF

(fig. 38)

1895 *Pristoceraea* KARSCH, Entomol. Nachr.Bl., Wien, **21**:349. – Species typica: *Agarista eriopis* Herrich-Schäffer (by original designation).
1901 *Mimeusemia* HAMPSON (partim), Cat. Lepidopt. Phal. Brit. Mus., **3**:607.
1955 *Charitosemia* KIRIAKOFF, Bull. Inst. r. Sci. nat. Belgique, Bruxelles, **31** (69):1. – Species typica: *Protoceraea geraldi* KIRBY (by original designation).

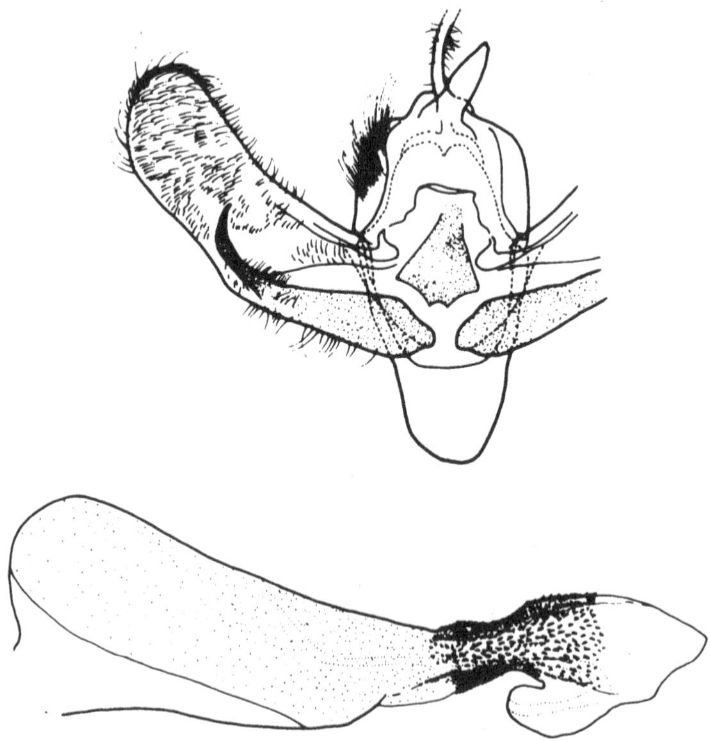

Fig. 38. *Charitosemia albigutta*. B. M. Noct. 8383. Type

D i a g n o s i s : Proboscis fully developed. Palpi nearly reaching the upper margin of the eye, porrect, first and second joints with moderate pilosity, third joint naked, as long as the first two together. Frons with a truncate conical prominence. Antennae of male and female filiform. Legs with moderate pilosity, denser on fore-legs in male; spurs of hind tibiae about of equal length. Abdomen with a basal crest only. Nervation: In fore wings, vein 2 given off from beyond three-quarters of the cell; 3, 4 and 5 well separated; areole narrow; 6 from the upper cell-angle; 7 and stalk of 8 + 9 from the extremity of the areole; 10 from the upper angle thereof. In hind wings, vein 2 issuing from the cell very near the angle; 3, 4 and 6, 7 from the cell-angles; discocellulars faintly incurved; 5 weak, from middle of the discocellulars.

Male genitalia: Uncus rather short, not longer than half the costa, very narrow, arched. Tubus analis short and rather narrow. Fold of vinculum narrow, bearing moderately long pilosity. Valva elongate, moderately broad; costa regularly incurved; termen broadly rounded; sacculus regularly outcurved; fold of sacculus rather narrow, reaching to middle of the valva, and ending in a short, arched harpe. Aedeagus very strongly developed, twice as long as the costa, very robust, narrowing distally, and covered subterminally and terminally with numerous spicule-shaped cornuti; fultura inferior with the proximal margin truncate-triangular, excised at middle, and narrowing distally. Saccus semi-elliptical.

D i s t r i b u t i o n : West, Central and East Africa.

Key to the species

1 Fore wings mostly reddish-brown with pale yellow markings; pectus orange *Ch. geraldi*, p. 117
1' Fore wings black with white markings; pectus black . . . . *Ch. albigutta*, p. 117

Charitosemia albigutta KARSCH comb. nova

1895 *Pristoceraea albigutta* KARSCH, Entomol. Nachr.Bl., Wien, **21**:357; pl. 2, fig. 2 – T e r r a t y p i c a : Coast of West Africa (Type, ♂, in Berlin Museum).
1901 *Mimeusemia albigutta* – HAMPSON, Cat. Lepidopt. Phal. Brit. Mus., **3**:609.

D i a g n o s i s : Head and thorax black; palpi and sides of frons marked with whitish; a dorsal whitish stripe on collar and thorax; pectus black; legs orange, marked with black; abdomen orange with black dorsal crests. Fore wings black with white markings, viz., a basal subcostal patch; a triangular medial spot on the subcostal vein; a rounded patch below cell; two postmedial spots between veins 2 and 8; a sinuous antemedial silver blue line; spot in cell-middle and a line on discocellulars also silvery blue, and a broken post-medial line. Hind wings orange; a blackish terminal band narrowing from apex to the anal angle and produced on the submedian fold. Length of fore wing 15 mm.

D i s t r i b u t i o n : Coast of West Africa. Specimens in the Rothschild collection (Brit. Mus.) labelled "Katanga" and "Angola" belong to the next species.

Charitosemia geraldi (KIRBY)

1896 *Protoceraea geraldi* KIRBY [ex errore pro *Pristoceraea*], Ann. Mag. nat. Hist., London, (6) **18**:377; pl. 19, Fig. 3. – T e r r a t y p i c a : Uganda, Narogaré (Type, ♂, in British Museum).

1901 *Mimeusemia geraldi* [as synonym of *M. albigutta*] – HAMPSON, Cat. Lepidopt. Phal. Brit. Mus., **3**:609.
1934 *Mimeusemia geraldi* – GAEDE in SEITZ, Großschmetterl. Erde, **15**:23.
1955 *Charitosemia geraldi* – KIRIAKOFF, Bull. Inst. r. Sci. nat. Belgique, Bruxells, **31** (69):1.

Diagnosis: Head and thorax black; second palpal joint pale yellow at the extremity; sides of frons marked with pale yellow; a pale yellow stripe on collar and thorax; pectus and legs orange; tibiae marked with black; tarsi black above; abdomen orange with black dorsal crests. Fore wings reddish chocolate brown, becoming blackish costally; markings pale yellow, viz.: a basal patch below costa; a triangular medial spot on the subcostal vein; a rounded patch edged with blackish below cell; rounded spots in postmedial area between veins 2–4 and 5–8; an antemedial sinuous silvery blue line; a silvery blue spot in cell, a line on the discocellulars, and short lines beyond the postmedial spots; cilia dark brown. Hind wings orange; a blackish brown terminal band, rather broad, extending on costa to before middle. Length of fore wing 15–18 mm.

Distribution: Uganda, Tanzania, southern Zaire (Lulua, Shaba).

Charitosemia geraldi geraldi (KIRBY)

Distribution: Uganda.

Charitosemia geraldi angustata (JOICEY & TALBOT)

1924 *Mimeusemia geraldi angustata* JOICEY & TALBOT, Bull. Hill Mus. Witley, **1**:555. – Terra typica: Marangu Plateau, Southwest Tanzania, 7000 ft. (Type,♂, in British Museum).

Diagnosis: Markings white instead of pale yellow; marginal band on hind wings narrower.

Distribution: Tanzania; Southern Zaire; Angola.

Genus **Ovios** WALKER

(fig. 39)

1855 *Ovios* WALKER, List Lepidopt. Het. Brit. Mus., 3:153. – Species typica: *Eudryas capensis* HERRICH-SCHÄFFER (by monotypy).

Diagnosis: Proboscis fully developed. Antennae bipectinate with rather short branches, still shorter in ♀, serrate towards base and apex. Palpi rather long, obliquely porrect, with the second joint fringed with hair in front, and the third moderate. Frons with a large truncate conical projection with a raised rim at the extremity. Thorax scaly, without crests; tegulae fringed with long rough hair. Tibiae fringed with long hair. Abdomen with dorsal crests on one or two proximal rings in ♂, on the first three rings in ♀. Nervation: in fore wings, vein 2 arising from about seven-elevenths of the cell; 3, 4, 5 approximated and equidistant; areole narrow; 6 from the upper cell-angle; 7 and the short stalk of 8 + 9 from the extremity of the areole; 10 from the upper side thereof, near extremity. In hind wings, vein 2 arising from two-thirds of cell; 3, 4 and 6, 7 from the cell-angles; discocellulars faintly in-

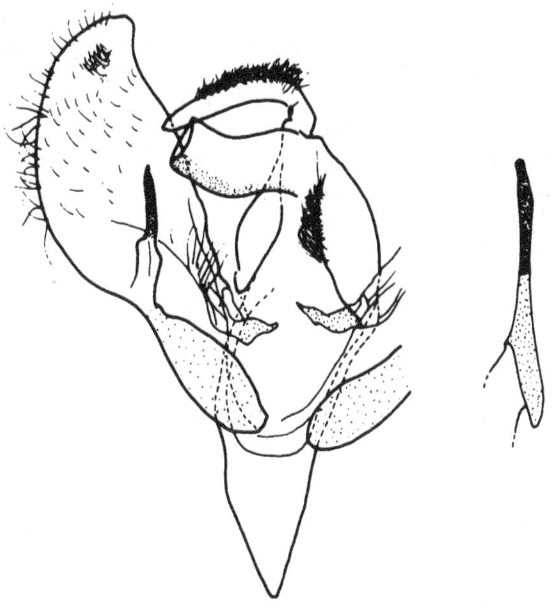

Fig. 39. *Ovios capensis* (HERRICH–SCHÄFFER). B. M. Noct. 8475

curved; 5 obsolescent, from middle of the discocellulars; 8 anastomosing at a point with cell near base.

Male genitalia: Uncus relatively small, somewhat longer than half of costa, compressed, arched at the base only. Tubus analis broad, sclerified ventrally. Fold of vinculum rather broad, fringed with hair. Valva leaf-shaped, slightly upcurved terminally; costa very faintly outcurved; sacculus faintly outcurved to end of fold, then much more distinctly so; fold elongate-elliptical, reaching to just one-third of sacculus, ending in an almost straight, broadish basally, slender terminally harpe. Aedeagus about three-quarters of costa in length, spoon-shaped proximally, then very slender, straight; vesica bearing a number of minute cornuti. Saccus elongate, triangular.

Distribution: South and East Africa.

Key to the species

1 Head and thorax white; fore wings pale yellowish green . . . . *O. capensis*, p. 119
– Head, thorax and fore wings pale grey brown . . . . *O. nealces*, p. 120

Ovios capensis (HERRICH-SCHÄFFER)

1853 *Eudryas capensis* HERRICH-SCHÄFFER, Außereurop. Schmetterl.:fig. 115–116. – Terra typica: Natal (Type, ♂, in British Museum).
1855 *Ovios capensis* – WALKER, List Lepidopt. Het. Brit. Mus., 3:753.

Diagnosis: Head and thorax white; palpi black, fringes of the second joint and the extremities of the second and the third orange; frons, vertex, occiput, collar, tegulae and thorax above marked with black; tegulae fringed with long orange hair; pectus and legs orange; fore tibiae and the tarsi banded black and white; abdomen orange with dorsal, lateral and sublateral series of black spots. Fore wings pale yellowish green; a black basal spot, edged with white, and a white patch above base of vein 1; costa sprinkled with black to beyond middle, expanding between the cell- and discocellular marks; a large wedge-shaped white patch in, just below and beyond cell, extending a as a fascia on costa to the apex; cell-mark round, deep red edged with yellowish and with a few black scales, and with at the centre some white scales; discocellular mark lunulate, deep red edged with yellowish, and partly with black, and with a white centre; a black point below costa near the apex; a broad terminal white band with a diffused lunulate black line on it before the termen, with red patches on it below the costa and above the dorsum, and a series of large terminal red lunules slightly edged with white; cilia whitish with a black medial line. Hind wings orange yellow with a large black discocellular lunule and a subterminal series of more or less triangular black spots. Length of fore wing 20–24 mm.

Distribution: South Africa.

Ovios capensis capensis (HERRICH-SCHÄFFER)

Distribution: Republic of South Africa: Cape Province, Natal.

Ovios capensis septentrionalis (HAMPSON)

1910 *Ovios septentrionalis* HAMPSON, Cat. Lepidopt. Phal. Brit. Mus., 9:427; pl. 146, fig. 8. – Terra typica: Mashonaland (Type, ♀, in British Museum).
1934 *Ovios capensis septentrionalis* – GAEDE in SEITZ, Großschmetterl. Erde, **15**:24.

Diagnosis: Collar and base of fore wings orange-red; discocellular spot rather V-shaped.

Distribution: Mashonaland.

Ovios nealces FAWCETT

1915 *Ovios nealces* FAWCETT, Proc. zool. Soc. London, **1915**:95; pl. 1, fig. 8. – Terra typica: Tanzania, Kedai (Types, ♂, ♀, in British Museum).

Diagnosis: Head, thorax and fore wings pale grey-brown with a violet reflection; abdomen and hind wings pale ochreous. Fore wings with a submarginal row of red-brown spots between veins, merging into a submedial band towards costa near the apex; a short white band from costa to vein 4 where it surrounds the discocellular mark; an indistinct round cell-spot; costa suffused with grey sprinkling. In the ♀ white band is broader and more diffused. Hind wings with a broad fuscous marginal band. Length of forewing 17–18 mm.

Distribution: Tanzania.

Genus **Leucovis** HAMPSON

(fig. 40)

1908 *Leucovis* HAMPSON, Cat. Lepidopt. Phal. Brit. Mus., 7:8. – Species typica: *Pristoceraea alba* ROTHSCHILD (by original designation).
1909 *Perimagia* STRAND, Dtsch. entomol. Z., Berlin, **1909**:36. – Species typica: *Pristoceraea alba* ROTHSCHILD (by original designation).

Diagnosis: Proboscis fully developed; antennae of ♂ minutely ciliated; palpi porrect, short, moderately fringed with hair below; frons with a large truncate conical projection, with raised rim at the extremity, and a corneous plate below it; thorax without crests; tibiae moderately fringed with hair; fore tibiae with tufts of hair on the joints above; abdomen with some rough hair at the base, but without crests. Nervation: in fore wings, vein 2 given off at about two-thirds of cell; 3, 4, 5 approximated, equidistant; discocellulars incurved; 6 from the lower cell-angle; areole small; 7 and the stalk of 8 + 9 from its extremity; 10 from its upper margin very near the extremity. In hind wings, vein 2 given off at five-sevenths of cell; 3, 4 and 6, 7 from the cell-angles; discocellulars well incurved; 5 obsolescent, from its middle; 8 anastomosing with cell a short while near the base.

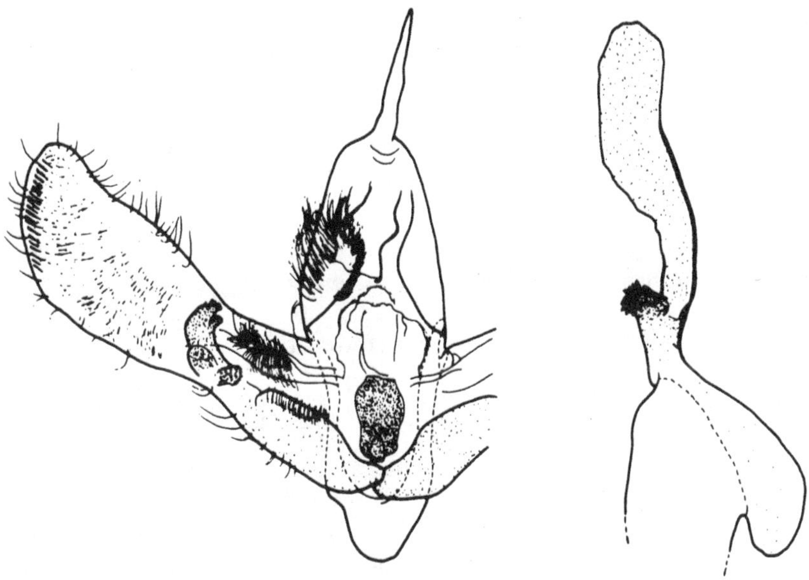

Fig. 40. *Leucovis alba* (ROTHSCHILD). B. M. Noct. 8480. Typus

Male genitalia: Uncus rather short, slightly less than one-half of costa, slender, pointed. Tubus analis moderately broad. Fold of vinculum broad at middle, with the pilosity rather abundant and placed along both the inner and the outer margin. Valva elongate and rather broad basally and terminally, narrowing just before middle; apex rounded; sacculus with a fringe of short stiff, spinelike hair along the terminal fifth; fold of sacculus about one-half of the breadth of valva at base, not reaching middle of sacculus, and bearing terminally a rather elaborate harpe, short, broad, curved and bearing at the extremity a small process and several spines; at the base a short spiny process. Aedeagus with a very long spoon-shaped basal, and a short, straight terminal portion, slightly more than one-quarter of the former, bearing distally a short club-shaped process with a number of spine-shaped cornuti on it; everted vesica longer than the aedeagus, with a lateral row of minute spinelike cornuti; fultura inferior elliptical. Saccus short, more or less semi-circular.

Distribution: South, Central and East Africa.

Key to the species

1 Fore wings with a narrow black costal and terminal margin . . . . ***L. alba***, p. 122
1' Fore wings with a broad blackish-brown apical and terminal band . . . . . . ***L. lepta***, p. 122

Leucovis alba (ROTHSCHILD)

1897 *Pristoceraea alba* ROTHSCHILD, Novit. Zool., **4**:183. – Terra typica: South Africa (Type, ♂, in British Museum).
1902 *Ovios alba* – HAMPSON, Ann. South Afr. Mus., Capetown, **2**:295.
1908 *Leucovis alba* – HAMPSON, Cat. Lepidopt. Phal. Brit. Mus., **7**:8.
1909 *Perimagia alba* – STRAND, Dtsch. entomol. Z., Berlin, **1909**:534–536.

Diagnosis: Head orange; palpi black, with the first and second joints black at the extremity; lower part of frons and antennae black; neck and shoulders black; thorax yellowish-white; pectus orange; fore and mid tibiae and tarsi black; abdomen orange, with the base yellowish-white. Fore wings yellowish-white; costa and termen narrowly black; cilia black mixed with some white. Hind wings white; termen with a slight black line between vein 6 and the submedian fold; cilia black between these points. Length of fore wing 19–22 mm.

Distribution: Rhodesia; Republic of South Africa (Transvaal, Mashonaland); Malawi; Zambia; Moçambique; Zaire (Shaba).

Leucovis lepta (FAWCETT)

1917 *Hylemera lepta* – FAWCETT, Proc. zool. Soc. London, **1917**:243; pl. 1, fig. 29. – Terra typica: Kenya, Kedai (Type, ♀, in British Museum).
1924 *Leucovis lepta* ROTHSCHILD, Ann. Mag. Nat. Hist., London, (9) **14**:315.
1924 *Leucovis latifascia* ROTHSCHILD, Ann. Mag. Nat. Hist., London, (9) **14**:315. – Terra typica: Kenya, Kibwezi (Type, ♀, in British Museum).

Diagnosis: Head and pectus yellow; thorax and urites 1 to 3 white above, yellow below; other urites yellow. Fore wings white; costa fuscous; a broad apical and terminal fuscous band, the proximal margin of which is angled outwardly between veins 3 and 4. Hind wings white, with a very narrow fuscous border, which is almost linear between veins 2 and 6. Length of fore wing 16 mm.

Distribution: Kenya.

Genus **Acantuerta** HAMPSON

(fig. 41)

1907 *Acantuerta* HAMPSON, Ann. Mag. nat. Hist., London, (7) **19**:241. – Species typica: *Tuerta thomensis* JORDAN (by original designation).

Diagnosis: Proboscis fully developed; antennae slightly dilated towards the extremity; palpi upturned, with the second joint fringed with long hair in front, and the third porrect, somewhat longer than in *Tuerta*, and flattened at the extremity; frons with a truncate conical prominence, with a raised rim in front, and a corneous plate below it, very narrow above between the eyes which are naked; mid and hind tibiae spined; hind tibiae with the spurs rather short; abdomen with dorsal crests on the i and ii urites. Venation: in fore wings, vein 2 from about three-quarters of cell; 3 and 4 well separated; 5 slightly nearer to 4 than 4 is to 3; discocellulars incurved; 6 from upper cell-angle; areole long, narrow; 7 and the moderately long stalk of 8 + 9 from tip of areole; 10 from upper side of areole, somewhat proximal. In hind wings, vein 2 from three-fifths of cell; 3, 4 and 6, 7 from the cell-angles; discocellulars incurved; 5 obsolescent, from middle of discocellulars.

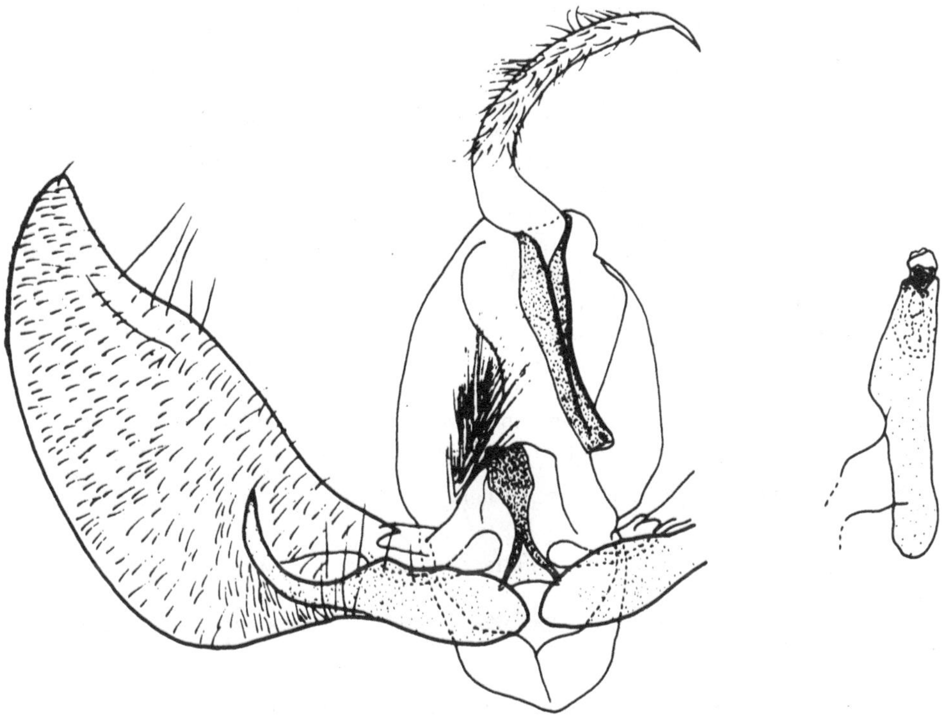

Fig. 41. *Acantuerta thomensis* JORDAN B. M. Noct. 8300

Male genitalia: Uncus about three-quarters of costa in length, shaped like a swan's neck, with a sharp terminal hook. Tubus analis narrow, well sclerified. Fold of vinculum broad, bearing on the proximal half a bunch of stiff hair. Valva leaf-shaped, with the base very narrow; costa slightly incurved in terminal third; apex subacute; sacculus outcurved, more distinctly along the proximal part of the broader portion of valva; fold of sacculus broad, elliptical, only reaching to one-fourth of sacculus, ending in a rather small sickle-shaped harpe. Aedeagus very small, about half of costa in length, rather slender but slightly more robust in distal half; vesica bearing terminally a patch of minute cornuti; fultura inferior narrow, shaped about like a wine-glass. Saccus short, rounded.

D i s t r i b u t i o n : Sao Thomé Island and the Lado Enclave.

Key to the species

1 Palpi, frons and legs marked with white; pectus and underside of abdomen whitish . . . *A. thomensis*, p. 124

1' Palpi, frons, legs, pectus and underside of abdomen without white . . . . . . *A. ladina*, p. 124

Acantuerta thomensis (JORDAN)

1904 *Tuerta thomensis* JORDAN, Novit. Zool., **11**:445. – T e r r a t y p i c a : St. Thomé Island (Types, ♂, ♀, in British Museum).
1907 *Acantuerta thomensis* – HAMPSON, Ann. Mag. nat. Hist., London, (7) **19**:241.
1913 *Aegocera thomensis* – JORDAN in SEITZ, Großschmetterl. Erde, **15**:19; pl. 4d.

D i a g n o s i s : Body dark chocolate brown; 2nd joint of palpi with the tip whitish in front; first joint of antennae and upper part of frons laterally whitish; pectus and femora whitish; dorsal abdominal crests with a metallic gloss; anal hair yellowish; underside of abdomen whitish. Fore wings deep chocolate brown, darker and tinged with purplish in female; a dull greenish fascia of diffused scales edged with chocolate above the media from near base to beyond middle, crossed by an oblique chocolate striga and dilated to near dorsum before middle; a small chocolate ring with greenish centre in cell-middle; a discocellular greenish bar edged with chocolate, and with a narrower band beyond it; postmedial line greenish, edged with chocolate; a wedge-shaped greenish patch on chocolate suffusion beyond it, from below costa to vein 4; a rounded patch in submedian interspace; a fine chocolate terminal line. Hind wings blackish brown with a reddish tinge; an orange patch in, below and beyond cell, smaller in female, and with brown nervural streaks in it; cilia whitish basally and at the extremities. Length of fore wings 18–21 mm.

D i s t r i b u t i o n : Sao Thomé Island.

Acantuerta ladina JORDAN

1926 *Acantuerta ladina* JORDAN, Novit. Zool., **33**:377. – T e r r a t y p i c a : Shambo (Type, ♂, in British Museum).

D i a g n o s i s : Body dark chocolate brown, without white markings on palpi, frons, legs and underside of abdomen; pectus dark brown; sides of rons rufous brown. Fore wings deep chocolate brown; markings as in *A. thomensis*, but the ring in cell replaced by an oblique line bearing some plumbeous scales; a plumbeous line behind the cell, at onefifth of dorsum; a greenish discocellular bar, bearing some plumbeous scales, as are the line connecting it to the dorsum, and the distal margin of the discal line. Hind wings orange; a black terminal band, 5 mm wide below the apex, and 3 mm before the tornus. Length of fore wing 23 mm.

D i s t r i b u t i o n : Lado Enclave (Shambo = terra typica).

B. Madagascan Genera

As pointed out in the Introduction to Part Two, the Agaristid fauna of Madagascar is completely endemic, there being not a single genus (with a single possible exception, dealt with elsewere) common to the African continent and the island. The dominant genus, *Rothia* WESTWOOD, comprising about two-thirds of the known insular species, certainly is of continental african origin, being probably an offshoot of the dominant Ethiopian genus *Heraclia* HÜBNER. Some of the remaining genera: *Pararothia* KIRIAKOFF & VIETTE, *Pristoceraea* KARSCH, *Schausiana* KIRIAKOFF & VIETTE, are more or less closely related to *Rothia. Antigodasa* KIRIAKOFF & VIETTE probably is less closely related to the above group on account of the curiously hypertrophied aedeagus. The origin of *Nesaegocera* KIRIAKOFF & VIETTE, confined to the Comores, may be directly continental. *Ancarista* JORDAN is unique in the absence of an harpe and in the presence of a large patch of strong spines in the distal portion of the valva; it might be related to the next two genera but for the absence of a stridulating organ in the male. Finally, *Musurgina* and *Pemphigostola*, which possess that structure, may well have an eastern origin, in opposition with the foregoing genera. The close affinity of those two genera with the genus *Saigonita* KIRIAKOFF from South Vietnam has been recently pointed out by the describer of the last genus.

The origin of the Madagascan Agaristid fauna may be considered accordingly, with a reasonable degree of probability, as being double, viz. Continental Ethiopian, and Oriental. The latter element is probably the more recent and limited presently to two or perhaps three genera.

Key to the genera

1 Fore tarsi much elongate in male, with the first joint thickened, and the second the longest . . . . **Schausiana**, p. 141
1′ Fore tarsi normal 2
2 Fore wings with a stridulating organ 3
2′ Fore wings without stridulating organ 4
3 Last palpal joint pointed; fore wings with veins 6–10 from the extremity of the areole **Pemphigostola**, p. 153
3′ Last palpal joint clubbed; fore wings with vein 6 from the cell-angle, 7–10 from the extremity of the areole . . . . **Musurgina**, p. 152
4 Fore, or mid- and hind tarsi in male, and tibiae bearing a long pilosity 5
4′ Tarsi without long pilosity 6
5 Fore tarsi in male densely pilose; hind wings with vein 2 from well before the angle of cell . . . . **Arctiopais**, p. 145
5′ Mid- and hind tarsi in male with long pilosity; Hind wings with vein 2 from very near the angle of cell . . . . **Pristoceraea**, p. 140
6 Last palpal joint short 7
6′ Last palpal joint long, or very long, porrect 8
7 Last palpal joint porrect, and slightly dilated terminally; hind wings with vein 5 obsolescent . . . . **Ancarista**, p. 150
7′ Last palpal joint hanging, and not dilated terminally; hind wings with vein 5 normal . . **Antigodasa**, p. 148
8 Last palpal joint clubbed; fore wings with vein 10 from the extremity of the areole, stalked with 8 + 9 . . . . **Nesaegocera**, p. 143
8′ Last palpal joint not clubbed; fore wings with vein 10 from the extremity of the areole, separate from the stalk of 8 + 9 9
9 In hind wings, veins 3, 4 and 6, 7 from a point, at the cell-angles . . . . **Rothia**, p. 126
9′ In hind wings, veins 3, 4 and 6, 7 stalked, at the cell-angles 10
10 Last palpal joint long, smooth, porrect . . . . **Arrothia**, p. 146
10′ Last palpal joint very short . . . . **Pararothia**, p. 138

Genus **Rothia** WESTWOOD

(fig. 42)

1877 *Rothia* WESTWOOD, Trans. Linn. Soc., Zool., (2) 1:204. – Species typica: *Agarista pales* GUÉRIN (by subsequent designation by Hampson 1901).

Diagnosis: Proboscis fully developed; palpi upturned; first and second joints with rather long hair in front; third joint well developed, smooth, mostly porrect; frons with a truncate conical prominence, with a slightly raised rim at the extremity; antennae distinctly dilated towards the extremity; abdomen in male with slight lateral tuft of hairscales behind the stigmata. Fore wings with the median cell mostly shorter in male, and often more elongate in that sex, with the termen strongly oblique. Nervation: in fore wings, vein 2 from three-quarters of the cell; 3 and 4 from the angle; 5 from the lower part of the discocellulars; upper discocellular angles inwards; areole narrow; 6 from the upper cell-angle; 7 and 8 + 9 from the extremity of the areole; 10 from the upper angle thereof. In hind wings, vein 2 from three-quarters of the cell; 3, 4 and 6, 7 from the cell-angles; discocellulars faintly incurved; 6 obsolescent, from middle of the discocellulars.

Male genitalia: Uncus very long, generally as long as or longer than the valva, compressed and arched, often with long and dense pilosity. Tubus analis elongate, sometimes as long as the uncus. Vinculum bearing the usual bunch of stiff hair. Valva typically shaped like an axe, with the base narrow, and with a rather broad

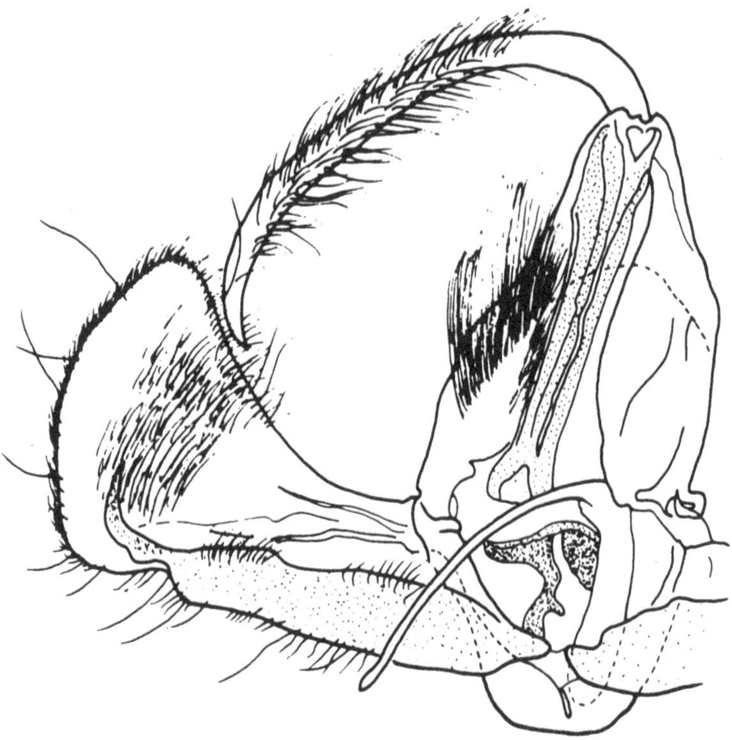

Fig. 42. *Rothia pales* (GUÉRIN–MÉNEVILLE). Prep. P. Viette 5021. Paris Museum

fold of the sacculus, and a short, arched harpe. Aedeagus rather short, bearing in most species a long and slender terminal process, generally more or less arched and placed at a right angle to the aedeagus, sometimes the aedeagus is bent at a right angle and without process, or slender, more or less straight, also without process; fultura inferior variable. Saccus short, rounded.

Key to the species

1 Hind wings red or orange above, with a dark border 2
1' Hind wings not red or orange above 5
2 Fore wings with several white or yellowish spots 3
2' Fore wings with a single post-cellular band, but sometimes two small cellular spots . . . 4
3 Hind wings orange yellow; size small (length of fore wing about 15 mm) . ***R. dayremi***, p. 129
3' Hind wings carmine red; size larger (length of fore wings 20 mm or more) . ***R. arrosa***, p. 130
4 Fore wings with two small cell-spots and some silvery-blue markings . . . . . ***R. holli***, p. 129
4' Fore wings without such spots and markings . . . . ***R. watersii***, p. 133
5 Underside of hind wings with a red, orange or yellow spot 6
5' Underside of hind wings without bright spot 19
6 Underside of hind wings with a basal carmine red or scarlet spot 7
6' Underside of hind wings with an orange or yellow spot 11
7 Fore wings with a yellow discal spot 8
7' Fore wings with the markings differently coloured or disposed 9
8 Underside of hind wings with the red occupying more than half of costa . . . . ***R. hypopyrrha***, p. 133
8' Underside of hind wings with the red occupying less than half of costa . . ***R. simyra***, p. 132
9 Hind wings uniformly black, unmarked . . . . ***R. zea***, p. 134
9' Hind wings with a bluish discal spot 10
10 Fore wings with two light spots in cell . . . . ***R. distigma***, p. 134
10' Fore wings without cell-spots . . . . ***R. cruenta***, p. 130
11 Upperside of hind wings with a yellow spot 12
11' Upperside of hind wings with a bluish spot 16
12 Underside of hind wings with the yellow spot reaching the base of the costa 13
12' Underside of hind wings with the yellow spot not reaching the base of the costa 15
13 Underside of hind wings with the base and a discal spot orange . . . . . . ***R. powelli***, p. 128
13' Underside of hind wings with the base and a discal spot orange, the latter partly white 14
14 Fore wings with a yellow cell-spot . . . . ***R. viossati***, p. 132
14' Fore wing without yellow cell-spot . . . . ***R. westwoodi***, p. 131
15 Upperside of hind wings with a yellow spot in the distal half of the wing . ***R. agrius***, p. 134
15' Upperside of hind wings with a yellow spot beginning at the basal fourth of the wing . ***R. pedasus***, p. 135
16 Fore wings with a white spot at the cell-end, and a series of white spots beyond . . . . ***R. pales***, p. 135
16' Fore wings with a light band or a diffuse spot immediately beyond the cell-end 17
17 Fore wings with a yellowish postcellular band . . . . ***R. micropales***, p. 127
17' Fore wings with a diffuse postcellular mark . . . . ***R. turlini***, p. 136
18 Upperside of hind wings with a white spot, sometimes with blue edges 19
18' Upperside of hind wings without discal spot 20
19 Upperside of hind wings with a yellowish discal spot, without blue edges . . . . ***R. hampsoni***, p. 131
19' Upperside of hind wings with a white discal spot, edged with blue . . . . . . ***R. lasti***, p. 137
20 Fore wings with an oblique yellow band beyond the cell . . . . ***R. nigrescens***, p. 137
20' Fore wing with a short band on the discocellulars, and a small postcellular spot . . . . . ***R. rhaco***, p. 130

Rothia micropales (BUTLER)

1879 *Rothia micropales* BUTLER, Ann. Mag. Nat. Hist., London, (5) **4**:235. – Terra typica: centr. Madagascar (Type, ♂, in British Museum).
1880 *Eusemia tranquilla* BUTLER, Ann. Mag. nat. Hist., London, (5) **5**:340. – Terra typica: Madagascar (Type, ♂, in British Museum).

1884 *Agarista epipales* MABILLE, C. R. Soc. entomol. Belgique, Bruxelles, **28**:189. – Terra typica: Madagascar (Type, ♀, in British Museum).
1892 *Anaphela tranquilla* – KIRBY, Syn. Cat. Lepidopt. Het.:25.
1892 *Anaphela epipales* – KIRBY, Syn. Cat. Lepidopt. Het.:25.
1901 *Rothia tranquilla* – HAMPSON, Cat. Lepidopt. Phal. Brit. Mus., **3**:582; pl. 53, fig. 19.
1913 *Rothia tranquilla* forma *tenuis* JORDAN in SEITZ, Großschmetterl. Erde, **15**:9. – Terra typica: Madagascar Center (Type, ♀, in British Museum).

Diagnosis: Head and thorax black, spotted with pale yellowish; pectus and legs orange; legs streaked with black; abdomen black above, orange laterally and on the underside; anal tuft black, with some orange hair at the extremity. Wings chocolate black. Fore wings without cell-spot; postdiscal band oblique, creamy-white, rather short, from below the subcostalis to interspace I, somewhat varying in shape; cilia white apically. Underside with the base ochreous orange. Hind wings with a large, or very large cream-coloured patch, edged with blue, below and beyond the lower cell-angle; cilia broadly white at the apex. Underside ochreous orange, with a broad black marginal band, occupying the three-quarters of the costa and broadly reaching the anal angle. Length of fore wing 20–21 mm.

Distribution: Madagascar.

Rothia micropales micropales (BUTLER)

1913 *Rothia micropales* ssp. *micropales* – JORDAN in SEITZ, Großschmetterl. Erde, **15**:9; pl. 3e.

Diagnosis: Discal band on fore wings reaching the anal nervure, pointed towards costa, otherwise uniformly broad (3 mm). Spot on hind wings about 4 mm in diameter.

Distribution: Central Madagascar.

Rothia micropales caecata JORDAN

1913 *Rothia micropales* ssp. *caecata* JORDAN in SEITZ, Großschmetterl. Erde, **15**:9. – Terra typica: Morondava (Type, 000, in British Museum).

Diagnosis: Discal band on fore wings not reaching the anal nervure, about $2^1/_2$ mm wide costally, and 4 mm elsewhere. Spot on hind wings about 8 mm in diameter, without the white centre. There also are some differences in the male genitalia, so that *R. micropales caecata* may well turn out to be a valid species.

Distribution: Southwestern Madagascar.

Rothia powelli (OBERTHÜR)

1909 *Anaphela powelli* OBERTHÜR, Étud. Lépidopt. comp., **3**:97; pl. 10, fig. 3. – Terra typica: Fianarantsoa (Type, ♀, in British Museum).
1913 *Rothia powelli* – JORDAN in SEITZ, Großschmetterl. Erde, **15**:11; pl. 5a.

Diagnosis: Head and thorax above black, marked with white; pectus yellow; abdomen orange with black edges to the urites; two last urites and anal tuft black. Fore wings reddish black; costal and in a lesser degree dorsal area sprinkled with silvery-blue; a white postmedial band, oblique, narrow (1 mm wide), nearly straight, running from the subcostal vein to vein 2; cilia concolorous. On the underside, base broadly orange, and the white band much broader. Hind wings reddish black; a broad discal white patch (7 mm wide), becoming distally metallic purplish blue; terminal black margin 4 mm wide before vein 5, and only 1 $^1/_2$ mm towards the anal angle. Underside orange, with a terminal black band. Length of fore wing 16 mm.

Distribution: Madagascar (Fianarantsoa).

Rothia dayremi (Oberthür)

1909 *Anaphela dayremi* OBERTHÜR, Étud. Lépidopt. comp., **3**:96; pl. 10, fig. 2. – Terra typica: Antsianaka (Type, ♂, in British Museum).
1913 *Rothia dayremi* – JORDAN in SEITZ, Großschmetterl. Erde, **15**:10; pl. 5 a.

Diagnosis: Head and thorax above reddish blackish brown; vertex and spots on the tegulae pale yellow; pectus and legs with orange pilosity; abdomen blackish brown with the urites margined with orange yellow; terminal tuft orange yellow laterally; underside orange yellow. Fore wings reddish blackish brown; a small pale yellow antemedial spot in cell; a dot below the cell; a pale yellow bar across the cell-extremity; pale yellow postmedial spots in interspace VI and between veins 3 and 5; cilia pale yellow in submedian fold. Underside with the discal area tinged with blackish. Hind wings orange yellow; base and a band above the anal margin dark reddish brown; anal margin orange yellow; terminal band dark reddish brown. Underside similar, but the brown marginal band broader. Length of fore wing 15 mm.

Distribution: Madagascar (Antsianaka).

Rothia holli (Oberthür)

1909 *Anaphela holli* OBERTHÜR, Étud. Lépidopt. comp., **3**:96; pl. 10, fig. 1. – Terra typica: Fianarantsoa (Type, ♂, in British Museum).
1913 *Rothia holli* – JORDAN in SEITZ, Großschmetterl. Erde, **15**:10; pl. 5 a.

Diagnosis: Head and thorax above black marked with white; tegulae with a white median spot; underside and legs orange; tarsi black; abdomen blackish brown, banded with orange. Wings blackish chocolate. Fore wings marked with silvery-blue, thus: at the base of the costa; a broken discocellular bar; base below the cell; angled line below the middle of the dorsum; a praetornal spot; two small triangular yellowish white spots in the cell; an oblique yellowish white postmedial band from below costa to vein 2; cilia blackish brown. Underside with the base orange. Hind wings deep orange red; base and a line before the anal margin, blackish; a chocolate black terminal band, occupying the whole of costa and reaching the anal angle; cilia blackish brown. Underside with the base orange. Length of fore wing 17 mm.

Distribution: Madagascar: Fianarantsoa.

Rothia cruenta JORDAN

1913 *Rothia cruenta* JORDAN in SEITZ, Großschmetterl. Erde, **15**:8. – Terra typica: Madagascar (Type, ♂, in British Museum).

Diagnosis: Head and thorax above black; first and second palpal joints, sides of frons, vertex, collar and tegulae spotted with white; base of the palpi with some orange; underside and legs orange; tarsi blackish; abdomen black, with fine whitish segmental edges; underside orange, barred with black towards the extremity; anal tuft black tipped with orange. Fore wings black; an oblique creamy white medial band from the upper cell-angle to vein 1, narrower costally, about 5 mm wide in median part; cilia white at the apex. Hind wings black, with an oblique elliptical blue patch with some white in centre, in and below the cell-extremity; cilia white apically. Underside with the base orange; discal area brick-orange; terminal band black. Length of fore wing 28 mm.

Distribution: Madagascar.

Rothia rhaeo (DRUCE)

1894 *Agarista rhaeo* DRUCE, Ann. Mag. Nat. Hist., London, (6) **14**:22. – Terra typica: Madagascar (Type, ♂, in British Museum).
1897 *Phalaenoides alluaudi* MABILLE, Ann. entomol. France, **1897**:219. – Terra typica: Madagascar (Type, ♀, in British Museum).
1901 *Rothia rhaeo* – HAMPSON, Cat. Lepidopt. Phal. Brit. Mus., **3**:584; pl. 54, fig. 16.
1901 *Rothia alluaudi* – HAMPSON, Cat. Lepidopt. Phal. Brit. Mus., **3**:584.
1912 *Rothia allaudi* STRAND (ex errore pro *alluaudi*), Lepidopt. Cat., **5**:18.
1913 *Rothia rhaeo* forma *alluaudi* – JORDAN in SEITZ, Großschmetterl. Erde, **15**:10.

Diagnosis: Head and thorax above black, marked with white; last palpal joint smooth, distinctly longer than in other species of *Rothia*; abdomen black with the extremity of the anal tuft yellow. Fore wings black, marked with white, viz.: a basal point; two minute triangular marks in the cell, one subbasal, the other in the distal half; two large squarish postmedial spots, generally connected proximally, one just below the costa, the other in interspaces I–III; cilia white at the apex and at the tornus. Hind wings black with three terminal white dots near the anal angle; cilia white at the apex and near the anal angle. Underside with a small squarish spot on the discocellulars. Length of fore wing 14–17 mm. *Rothia alluaudi* Mabille, often considered as a good species, does in no way differ from *R. rhaeo* proper, even not as a "form".

Distribution: Madagascar.

Rothia arrosa JORDAN

1926 *Rothia arrosa* JORDAN, Novit. Zool., **33**:373. – Terra typica: Lake Alabatra (Type, ♀, in British Museum).

Diagnosis: Sexes slightly different. Male: head and thorax above black spotted with white; collar black, edged with white; tegulae sprinkled with white; underside of thorax and legs orange; tarsi ringed with yellowish and black; abdomen with some yellowish hair at the base; rest orange red, ringed with yellowish; underside spotted with black; anal tuft orange, with the base and sides black. Fore wings black, spotted with white, viz.: a basal dot; a triangular spot at cell-base; transversal cell-spot near cell-extremity; a rather indistinct spot in interspace I before middle; two large rounded spots beyond cell-end, one just below the subcostal vein, the other more distally in interspaces (I), II and III; a subterminal oval spot in interval I; cilia broadly spotted with white. Underside with the basal area orange. Hind wings minium red; extreme base black; a white dot at the base of cell; another, larger, on the discocellulars; a black marginal band, occupying the distal half of costa, 7 mm wide at the apex, narrowing towards the anal angle; cilia broadly spotted with white. Underside with the costal area orange. Female: abdomen with small black dorsal spots on the three proximal urites. Fore wings with silvery-blue markings on costa, at the base, between the cell-spots and on the discocellulars; costa with two basal white points; spot in interspace II much reduced. Hind wings with the marginal band broader; the white dots absent on the upperside. Length of fore wing 19–21 mm.

Distribution: Madagascar.

Rothia hampsoni OBERTHÜR

1916 *Rothia hampsoni* OBERTHÜR, Étud. Lépidopt. comp., **11**:247; pl. 370, fig. 3095. – Terra typica: Antsianaka (Type, ♀, in British Museum).

Diagnosis: Sexes not different. Head and thorax above black, marked with white; abdomen black above, orange laterally and on the underside; anal tuft black with the base orange, and a small white dorso-terminal dot. Fore wings black; an oblique yellowish-white postmedial band, running from the subcostalis to vein 2, narrower in the costal third, about 2 mm wide; cilia concolorous. Underside with the base orange. Hind wings black; a large white or yellowich-white patch from near costa to the anal angle, 7 mm wide; anal margin broadly black; cilia concolorous. Underside with the base orange, and the white patch reduced. Length of fore wing 14–18 mm.

Distribution: Madagascar.

Rothia westwoodi BUTLER

1879 *Rothia westwoodi* BUTLER, Ann. Mag. nat. Hist., London, (5) **4**:235. – Terra typica: Tananarive (Type, ♂, in British Museum).
1859 *Eusemia virguncula* MABILLE, Bull. Soc. philom. Paris, (7) **3**:136. – Terra typica: Fianarantsoa (Type, 000, in British Museum).
1892 *Anaphela westwoodi* – KIRBY, Syn. Cat. Lepidopt. Het.:25.
1892 *Anaphela virguncula* – KIRBY, Syn. Cat. Lepidopt. Het.:25.
1901 *Rothia vestvoodi* HAMPSON (ex errore pro *westwoodi*), Cat. Lepidopt. Phal. Brit. Mus., **3**:380; pl. 53, fig. 5.
1901 *Rothia virguncula* – HAMPSON, Cat. Lepidopt. Phal. Brit. Mus., **3**:580; pl. 53, fig. 7.

Diagnosis: Body and wings black; base and extremity of the second palpal joint, sides of frons and vertex, spotted with whitish or pale yellow; underside of thorax and legs orange; abdomen with fine segmental yellow lines; underside orange, with the penultimate segment spotted with black, and the extremity black. Fore wings with an oblique postmedial pale yellow band, running from below the costa to interspace I, about 2 mm wide, dilated in interspace IV. Underside with the base orange. Hind wings with a squarish pale yellow patch, occupying about the proximal half of costa, and running to middle of vein 2, excurved between veins 3 and 5, about 7 mm wide; pilosity of the anal margin orange. Underside with the base orange. Length of fore wing 16–18 mm.

Distribution: Madagascar.

Rothia viossati KIRIAKOFF & VIETTE

1973 *Rothia viossati* KIRIAKOFF & VIETTE, Bull. Soc. entomol. France, Paris, 78:217. – Terra typica: Vanjamanitra Forest (Type, ♂, in Paris Museum).

Diagnosis: Head and thorax above black marked with white; thorax below and legs orange; tarsi black; abdomen black above, the urites narrowly edged with yellowish; sides and underside yellowish, the latter tinged with orange, and with the terminal segments marked with black; anal tuft black tipped with yellowish. Wings deep black. Fore wings with a subbasal light yellow triangular spot occupying the proximal third of the cell, and a narrow stroke below the cell; an oblique yellow postmedial band, from below costa to vein 2, narrow costally, broader and elliptical between veins 5 and 2; cilia concolorous. Underside with the base orange. Hind wings with a large white discal spot, broadest on vein 4 (6 mm); cilia concolorous. Underside with the base broadly orange, that colour occupying also the anal region and the proximal part of the white spot. Length of fore wing 15 mm.

Distribution: Madagascar.

Rothia simyra WESTWOOD

1877 *Rothia simyra* WESTWOOD, Trans. Linn. Soc., Zool., (2) 1:204. – Terra typica: Madagascar (Type, ♂, in British Museum).
1892 *Anaphela simyra* – KIRBY, Syn. Cat. Lepidopt. Het.:25.
1913 *Rothia simyra* forma *nigrifimbriata* JORDAN in SEITZ, Großschmetterl. Erde, 15:8. – Terra typica: Morondava, Madagascar W (Type, ♂, in British Museum).

Diagnosis: Body and wings blacksh brown; second and third palpal joints and vertex spotted with yellowish white; sides of frons yellowish; collar, shoulders and tegulae spotted with yellowish white; underside of thorax and legs orange; tarsi blackish; abdomen with lateral triangular yellow spots; underside orange; anal tuft black, tipped with orange. Fore wings with a slightly oblique yellow discal band from just below costa to above dorsum, 3 mm wide, slightly expanding dorsally; cilia sometimes concolorous (forma *nigrifimbriata*), but mostly white at the apex. Hind wings black, unmarked; pilosity of the anal margin orange; cilia as in fore wings. Underside with a reddish orange patch becoming orange at the base and on the costal area, extending from costa of which it occupies the proximal quarter, to the anal margin just before the tornus. Length of fore wing 26 mm.

Distribution: Madagascar.

Rothia watersii (BUTLER)

1884 *Eusemia watersii* BUTLER, Ann. Mag., nat. Hist., London, (5) **14**:409. – Terra typica: Betsileo (Type, ♂, in British Museum).
1892 *Anaphela watersii* – KIRBY, Syn. Cat. Lepidopt. Het.:24.
1901 *Rothia vatersi* HAMPSON (ex errore pro *watersii*), Cat. Lepidopt. Phal. Brit. Mus., **3**:578; pl. 53, fig. 6.
1911 *Anaphela fianarantsoa* OBERTHÜR, Ann. Soc. entomol. France, Paris, **79**:471; pl. 16, fig. 5. – Terra typica: Fianarantsoa (Type, ♀, in British Museum).
1912 *Rothia watersi* – STRAND (ex errore pro *watersii*), Lepidopt. Cat., **5**:20.
1913 *Rothia fianarantsoa* – JORDAN in SEITZ, Großschmetterl. Erde, **15**:10.
1916 *Rothia watersi* var. *lutescens* OBERTHÜR, Étud. Lépidopt. comp., **2**:247. – Terra typica: Fianarantsoa, Madagascar Center (Type, ♂, in British Museum).

Diagnosis: Head and thorax above black, spotted with whitish; thorax below and pilosity of legs orange; tarsi black above, orange below; abdomen orange, with black segmental lines; basal and terminal urites black; anal tuft black tipped with orange; abdomen sometimes black ringed with orange. Fore wings black; a yellow dot below base of costa; an oblique postmedial pale yellow band, from below the costa to vein 1, 2–2 $^1/_2$ mm wide; cilia white apically. Underside with the band pure white. Hind wings orange red or golden yellow (*lutescens* OBERTHÜR); base black; terminal area black, broadening towards the costa of which it occupies the terminal half. Underside with no black at the base. In female (*R. fianarantsoa* OBERTHÜR), the abdomen above has more of orange, and the black marginal band on hind wings occupies the whole of the costa. Length of fore wing 20–25 mm.

Distribution: Madagascar.

Rothia hypopyrrha (BUTLER)

1878 *Eusemia hypopyrrha* BUTLER, Cist. Entomol., **2**:297. – Terra typica: A forest near Antananarive (Type, ♀, in British Museum).
1892 *Episteme hypopyrrha* – KIRBY, Syn. Cat. Lepidopt. Het.:26.
1901 *Rothia hypopyrrha* – HAMPSON, Cat. Lepidopt. Phal. Brit. Mus., **3**:578; pl. 53, fig. 8.

Diagnosis: Male: fore wings relatively narrow, with termen straight and very oblique. Body and wings black; head and thorax above with small yellow spots; thorax below and legs orange; tarsi mostly black; abdomen ringed with orange above; underside orange; anal tuft black tipped with orange. Fore wings with two small yellow spots below base of costa; a broad, 5–6 mm wide, discal yellow band from below costa to just above dorsum beyond the middle, occupying the discocellular region; cilia white apically. Underside with base and dorsal area orange. Hind wings black, with reddish pilosity on the anal margin; cilia white apically. Underside red orange, with a terminal blackish brown band, broader costally (7–8 mm), and narrowing to a point at tornus. Fore wings in female broader, with the termen less oblique, and slightly outcurved; otherwise not different. Length of fore wing 30 mm.

Distribution: Madagascar.

Rothia zea (HERRICH-SCHÄFFER)

1853 *Agarista zea* HERRICH-SCHÄFFER, Außereurop. Schmetterl.: fig. 34 and 35. – Terra typica: "Casamanca" (Type, ♂, in British Museum).
1874 *Eusemia zea* – BOISDUVAL, Rev. Mag. Zool., Paris, (3) 2:99.
1892 *Anaphela zea* – KIRBY, Syn. Cat. Lepidopt. Het.: 24.
1901 *Rothia zea* – HAMPSON, Cat. Lepidopt. Phal. Brit. Mus., 3:579 (♂, ♀).
1913 *Rothia zea* forma *zea* – JORDAN in SEITZ, Großschmetterl. Erde, 15:8.
1913 *Rothia zea* forma *divisa* JORDAN in SEITZ, Großschmetterl. Erde, 15:8. – Terra typica: Morondava, Madagascar W (Type, ♂, in British Museum).

Diagnosis: Head and thorax above black, marked with pale yellow; thorax below and legs orange; tibiae and tarsi blackish; abdomen black, spotted with orange laterally; underside orange with black bands; anal tuft orange with the base black. Fore wings chocolate black; two small pale yellow spots below the base of the costa; two spots in cell, one antemedial, the other in cell-end; a postcellular yellowish band between the subcostalis and vein 2, sometimes broken at middle to form two spots (forma *divisa*); cilia spotted with white, but entirely white apically and at the tornus. Underside with the base yellow orange. Hind wings uniform chocolate black; cilia chequered white and black, the apical and preanal spots of white being the largest. In the female form *divisa* the white spots are larger. Underside light scarlet orange, with a black terminal band; base paler, more pink in the form *divisa*. Lenght of fore wing 24--31 mm.

Distribution: Madagascar. The type specimen in the British Museum bears a label "Casamanca" (recte Casamance, Senegal). The latter being a West African locality, the mention is certainly erroneous.

Rothia distigma (MABILLE)

1897 *Agarista distigma* MABILLE, Ann. Soc. entomol. France, Paris, 1897:219. – Terra typica: Madagascar (Type, ♀, coll. OBERTHÜR; lost).
1901 *Rothia distigma* – HAMPSON, Cat. Lepidopt. Phal. Brit. Mus., 3:579; pl. 53, fig. 12.

Diagnosis: Fore wings more elongate in male, with termen oblique. Body and wings black; head and thorax above marked with pale yellow; pectus and legs orange; mid- and hind tibiae black; abdomen with fine segmental lines and lateral spots; underside orange; anal tuft black tipped with orange. Fore wings with about three yellow points below base of the costa; a small bluish white antemedial spot in cell, and another near the cell-extremity; a postmedial series of four bluish white spots, between veins 3 and 7, more distinct in female; some silvery blue scales at the base, and in cell; cilia white apically. Hind wings with a large rounded metallic greenish blue spot on the disc; cilia white at the apex. Underside pinkish orange, with a black terminal band, occupying the distal three-quarters of costa and reaching the anal angle. Length of fore wing 28–32 mm.

Distribution: Madagascar.

Rothia agrius (HERRICH-SCHÄFFER)

1853 *Agarista agrius* HERRICH-SCHÄFFER, Außereurop. Schmetterl.: fig. 33. – Terra typica: Madagascar (Type, 000, in British Museum).
1874 *Eusemia agrius* – BOISDUVAL, Rev. Mag. Zool., Paris, (3) 2:97.

1892 *Anaphela agrius* – KIRBY, Syn. Cat. Lepidopt. Het.:25.
1896 *Rothia simplex* ROTHSCHILD, Novit. Zool., 3:44; pl. 14, fig. 18. – Terra typica: Morondava (Type, ♀, in British Museum).
1901 *Rothia agrius* – HAMPSON, Cat. Lepidopt. Phal. Brit. Mus., 3:581.
1901 *Rothia pedasus* – HAMPSON (*simplex* synonyme of), Cat. Lepidopt. Phal. Brit. Mus., 3:581.

Diagnosis: Head and thorax above black marked with white; underside and legs orange; tibiae sprinkled with black; tibiae black; abdomen black with fine yellow orange segmental edges, broader laterally; underside orange except the two terminal urites. Fore wings black, more elongate in male, and with the termen much more oblique; a discal yellowish white band, becoming white towards tornus, about 3 mm wide, from just below costa to vein 1 near the tornus; cilia white apically. Underside not different. Hind wings black with a broad discal spot, yellowish white in male and the form *simplex*, and pure white in female of the typical form; cilia finely white apically. Underside with the discal spot reaching the base. In the form *simplex*, there is much more yellow on the abdomen above. Length of fore wing 27–30 mm.

Distribution: Madagascar.

Rothia pedasus (HERRICH-SCHÄFFER)

1853 *Agarista pedasus* HERRICH-SCHÄFFER, Außereurop. Schmetterl.: fig. 32. - Terra typica: Madagascar (Type, ♂, in British Museum).
1874 *Eusemia pedasus* – BOISDUVAL, Rev. Mag. Zool., Paris, (3) 2:98.
1892 *Anaphela pedasus* – KIRBY, Syn. Cat. Lepidopt. Het.:25.
1901 *Rothia pedasus* – HAMPSON, Cat. Lepidopt. Phal. Brit. Mus., 3:581.

Diagnosis: Head and thorax above black, spotted with pale yellow; base of palpi orange; pectus and legs orange; tibiae and tarsi streaked with black; abdomen black; basal crest orange; segments finely edged with orange and sprinkled with that colour, except at base and at middle; sides and underside light orange, with a series of black dots in the proximal half; anal tuft orange, with the base black. Fore wings blackish brown; an yellowish white basal dot; a subbasal cell-spot; a wedge-shaped spot near the cell-extremity; an oblique postcellular band from the subcostalis to vein 2, expanding below, about 4 mm wide; cilia white apically. Underside with the base orange. Hind wings blackish-brown, with the base browner, and a large pale yellowish patch from below costa to the anal margin, with some blue scales on it on the inner area, maximal width of the dark marginal band 11 mm; cilia white apically. Underside with the base yellow orange. Length of fore wing 25--28 mm.

Distribution: Madagascar.

Rothia pales (GUÉRIN)

1844 *Agarista pales* GUÉRIN, Iconogr. Règne Animal, Ins.:493; pl. 83, fig. 1. – Terra typica: Madagascar (Type, ♀, in British Museum).
1874 *Eusemia pales* – BOISDUVAL, Rev. Mag. Zool., Paris, (3) 2:100.
1892 *Anaphela pales* – KIRBY, Syn. Cat. Lepidopt. Het.:25.
1901 *Rothia pales* – HAMPSON, Cat. Lepidopt. Phal. Brit. Mus., 3:582; fig. 249.

D i a g n o s i s : Head and thorax above black, spotted with yellowish white; underside and legs orange; tarsi black above; abdomen black; segmental lines orange; underside orange, with the last four urites black; anal tuft black, with a few orange hair at the extremity. Fore wings elongate and rather narrow in male, with the termen oblique; black, with a whitish spot near the cell-extremity; an oblique post-medial band between veins 7 and 2, divided into spots by the black veins; spot in interspace IV more elongate; cilia tipped with white. Underside with the base ochreous yellow. Hind wings black, with a large rounded metallic blue patch in disc, forming sometimes a separate elongate spot in the submedian fold; cilia white apically. Underside with the basal area bright ochreous yellow; the black terminal band occupying the distal two-thirds of the costa and reaching the anal angle. Length of fore wing 28 mm.

D i s t r i b u t i o n : Madagascar.

Rothia metagrius (BUTLER)

1880 *Eusemia metagrius* BUTLER, Ann. Mag. nat. Hist., London, (5) **5**:25. – T e r r a t y p i c a : Madagascar (Type, in British Museum).
1892 *Anaphela virguncula (metagrius* synonym of) – KIRBY, Syn. Cat. Lepidopt. Het.:25.
1901 *Rothia virguncula* (*metagrius* synonym of) – HAMPSON, Cat. Lepidopt. Phal. Brit. Mus., **3**:580.

D i a g n o s i s : Head and thorax above black, marked with white; abdomen black above, with orange edges to the segments, broader laterally; underside orange; anal tuft black with an orange median spot. Wings black. Fore wings with the basal part of costa spotted with yellowish white and metallic blue; a yellowish white discal band, from just below the costa to the anal vein before the tornus, more or less straight, with a small notch at the lower cell-angle, and a corresponding very small distal tooth; cilia white at the apex. Underside with the base orange, and the median band broader. Hind wings with a transverse yellowish white discal patch varying in size, rounded distally, somewhat produced and sprinkled with black in the anal fold; cilia uniformly black. Underside yellowish white, with a broad black terminal band; extreme base, costal and anal margins pale orange. Length of fore wing 23 mm.

D i s t r i b u t i o n : Madagascar.

Rothia turlini KIRIAKOFF & VIETTE

1913 *Rothia tranquilla* forma *sinefascia* JORDAN, in SEITZ, Großschmett. Erde, **15**:9. – T e r r a t y p i c a : Not designated (Type lost).
1973 *Rothia turlini* KIRIAKOFF & VIETTE, Bull. Soc. entomol. France, Paris, **78**:216. – T e r r a t y p i c a : La Mandraka (Type, ♂, in Paris Museum).

D i a g n o s i s : Head and thorax above black, marked with white; thorax below and legs orange; tarsi black; abdomen orange above, black below, with lateral orange spots. Wings deep black. Fore wings with a diffuse discal band of white sprinkling, somewhat varying in size, but occupying most of the discocellulars and the bases of the interspaces II and III; cilia black. Underside with the base orange, and the discal band distinct, whole, from below the costa to interspace II. Hind wings bearing a blue discal patch with a whitish centre, 5 mm wide; anal pilosity orange. Underside orange with a black terminal band, 5 mm wide at the costa. Length of fore wing 20 mm.

Distribution: Central and eastern Madagascar.

Rothia lasti ROTHSCHILD

1896 *Rothia lasti* ROTHSCHILD, Novit. Zool., London, **3**:45. – Terra typica: Morondava (Type, ♂, in British Museum).
1913 *Rothia lasti* ssp. *lasti* – JORDAN in SEITZ, Großschmetterl. Erde, **15**:10; pl. 3b.
1913 *Rothia lasti* ssp. *epiera* JORDAN in SEITZ, Großschmetterl. Erde, **15**:10. – Terra typica: Madagascar (Type, ♂, in British Museum).

Diagnosis: Head and thorax above black, spotted with yellowish white; pectus and legs orange; part of tibiae and tarsi black; abdomen black; a lateral line, basal hair and underside orange. Wings black. Fore wings with a white spot at the cell-base, and another near the discocellulars; a postmedial band of conjoined spots between veins 7 and 2, narrower in the costal area; two costal spots placed at a right angle to the costa, the third parallel with the dorsum. Hind wings with a large white or bluish white discal patch of conjoined spots in interspaces I to VI, rather diffuse distally; cilia sometimes white apically. Underside not different, without any trace of orange. Length of fore wing 18,5–21 mm. – *R. lasti* ssp. *epiera* does not show any difference with the "typical" form.

Distribution: Madagascar. The type locality of ssp. *epiera* is stated to be Diego Suarez, but the type specimen in the British Museum has no locality.

Rothia nigrescens ROTHSCHILD

1896 *Rothia nigrescens* ROTHSCHILD, Novit. Zool., **3**:45; pl. 15, fig. 32. – Terra typica: Morondava (Type, ♂, in British Museum).
1923 *Rothia martha* OBERTHÜR, Étud. Lépidopt. comp., **21** (2):148; pl. 2, pl. 576, fig. 4890. – Terra typica: Sud du Madagascar (Type, ♀, in British Museum).

Diagnosis: Male: Head and thorax above black, spotted with pale yellow; pectus and legs orange; fore tibiae above and tarsi black; abdomen black above, and orange below. Wings black. Fore wings with some silvery blue scales on basal area below the costa, above vein 1 and in middle of the cell; an oblique pale yellow band just beyond the cell, from below costa to vein 1, with irregular edges; an antemedial dot in cell. Underside with the base orange. Hind wings unmarked; cilia spotted pale yellowish and black. Underside with the base orange. Female (*R. martha* OBERTHÜR): fore wings with two creamy white basal dots, otherwise not different. Length of fore wing 27 mm.

Distribution: Madagascar.

Genus **Pararothia** KIRIAKOFF & VIETTE
(fig. 43)

1974 *Pararothia* KIRIAKOFF & VIETTE, Faune Madagascar, Lepidopt. Agaristidae, **39**. – Species typica: *Rothia gracilis* JORDAN (designated by KIRIAKOFF & VIETTE 1974).

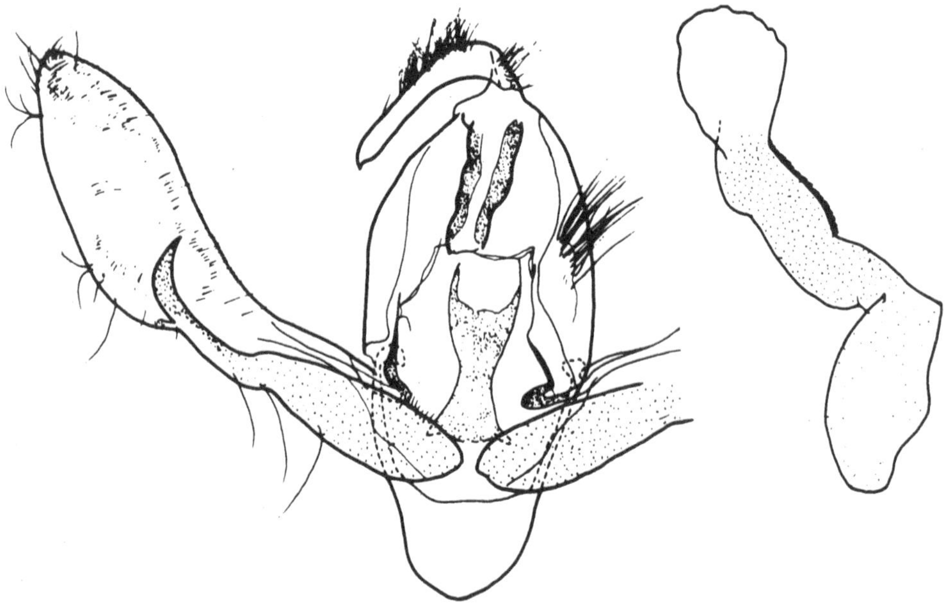

Fig. 43. *Pararothia gracilis* (JORDAN). B. M. Noct. 8515. Type

Diagnosis: Proboscis fully developed; palpi upturned; third joint very short; frons with a truncate conical projection, with raised rim at the extremity; antennae distinctly dilated towards the extremity; hind tibiae with two pair of spurs. Venation: in fore wings, vein 2 from beyond three-quarters of cell; 3 and 4 separate; 5 from the discocellulars near the angle; 6 from the lower cell-angle; areole narrow; 7, 8 + 9 and 10 from the extremity of the areole; common stalk of 8 + 9 very short. In hind wings, vein 2 from very near the extremity of the cell; 3, 4 and 6, 7 stalked; 5 very weak, from middle of the discocellulars.

Male genitalia: Uncus relatively short, not longer than half of costa. Tubus analis generally not longer than the uncus. Valva elongate, rather narrow, mostly slightly broader terminally; harpe short, placed rather distally. Aedeagus rather variable, typically robust, as long as the valva, but sometimes shorter and much expanded terminally; fultura inferior elongate, narrower in the middle and deeply excised distally. Saccus short, rounded.

Key to the species

1 Underside of hind wings orange at base, otherwise brownish carmine . . . *P. camilla*, p. 139
1' Underside of hind wings not so coloured 2
2 Underside of hind wings with base and anal margin orange . . . . *P. gracilis*, p. 139
2' Underside of hind wings with the proximal area orange, that colour occupying most of the discal spot . . . . *P. vieui*, p. 139

Pararothia gracilis (JORDAN)

1913 *Rothia gracilis* JORDAN in SEITZ, Großschmetterl. Erde, **15**:10. – Terra typica: Ambre (Type, ♂, in British Museum).
1974 *Pararothia gracilis* – KIRIAKOFF & VIETTE, Faune Madagascar, Lepidopt. Agaristidae, **39**

Diagnosis: Head and thorax above deep chocolate brown; first and second joints of palpi, sides of frons, space behind the antennae, collar and tegulae, spotted with white; thorax below and legs orange; tarsi black ringed with white; abdomen blackish brown; underside orange with some blackish towards the extremity. Fore wings deep chocolate brown; minute white subbasal, antemedian and median spots in the upper part of the cell; a broad oblique creamy white band from below the costa just beyond the cell to above the tornus; some silvery blue scaling below the basal part of the costa; oblique silvery blue line in middle of the cell, a curved discocellular line, and an oblique row of spots below the cell-end before the white band, with another line before the spot above the dorsum; a subterminal row of small silvery-blue spots from below the costa to vein 2; cilia dark brown. Hind wings blackish brown, slightly tinged with chocolate; a large creamy white spot beyond the cell between vein 1 and discal fold; cilia white at the extremity at the apex and at the anal angle. Underside orange at the base, and along the anal margin only. Length of fore wing 20 mm.

Distribution: Madagascar.

Pararothia camilla (OBERTHÜR)

1916 *Rothia camilla* OBERTHÜR, Étud. Lépidopt. comp., **21** (2):148–149; pl. 66, fig. 4886. – Terra typica: Southern Madagascar (Type, ♂, in British Museum).
1974 *Pararothia camilla* – KIRIAKOFF & VIETTE, Faune Madagascar, Lepidopt. Agaristidae, **39**: 73; fig. 60–61, 62.

Diagnosis: Head and thorax above black marked with white; thorax below and legs orange; tarsi black. Fore wings brownish chocolate red; costal area sprinkled with silvery blue; silvery spots between the cell-spots, forming further an inner line of three spots; an horizontal V on the discocellulars; a subterminal row of streaklets; a white basal point; two small cell-spots; a rather broad post-medial band from just below the costa to near tornus. Hind wings dark reddish brown; an elongate white spot in interspaces ii and iii, and at the base of interspace iv; another, smaller, placed more distally and slightly diffuse proximally; cilia white at the apex. Underside orange at the base; rest brownish carmine. Length of fore wing 18 mm.

Distribution: South Madagascar.

Pararothia vieui (VIETTE)

1966 *Rothia vieui* VIETTE, Lambillionea, **65** (9–10):88–89; fig. 2. – Terra typica: Forêt d'Analamazoatra, Station Perinet (Types, ♂, ♀, in Paris Museum).
1974 *Pararothia vieui* – KIRIAKOFF & VIETTE, Faune Madagascar, Lepidopt. Agaristidae, **39**: 77; fig. 63–65.

Diagnosis: Head and thorax above black marked with white; tegulae with a few orange hairscales along the inner margin; underside and legs orange; tarsi black. Fore wings reddish black, with some silvery blue markings, viz.: on the discocellulars; on the dorsum; a series of internervural streaks; a postmedial white band from the subcostalis to vein 2, slightly narrower in the middle; cilia white at the apex. Underside with the base orange, and with a terminal series of blue spots. Hind wings pure black; a large white discal spot, 6 mm wide, with the distal margin wavy; cilia white at the apex. Underside broadly orange in the basal area, that colour occupying two-thirds of the discal spot; a series of terminal bluish spots. The ♀ has blue costal sprinkling, and the postmedial band more regularly elliptical. Hind wings with the white band broader (10 mm), and on the underside the orange colour more restricted. Length of fore wing 18–23 mm.

Distribution: Madagascar.

Genus **Pristoceraea** KARSCH

(fig. 44)

1895 *Pristoceraea* KARSCH, Entomol. Nachr. Bl., Wien, **21** : 349. – Species typica: *Agarista eriopis* HERRICH-SCHÄFFER (by original designation).

1974 *Eriopana* KIRIAKOFF, Faune Madagascar. Lepidopt. Agaristidae **39**: 80. – Species typica: *Agarista eriopis* HERRICH-SCHÄFFER (by original designation).

Diagnosis: Proboscis fully developed; palpi with the first two segments moderately clothed with hair, the third segment porrect, slender, long and pointed; frons with a truncate conical prominence, with the terminal rim raised; antennae filiform in noth sexes; leg pilosity moderate; in ♂, pilosity of the forelegs denser and extending to the tarsi; proximal and distal spurs of the hind tibiae of equal length. Nervation: in fore wing, vein 2 from beyond three-quarters of the cell; 3, 4, 5 well separated, 6 from the lower cell-angle; areole narrow; 7 and 8 + 9 from the extrem-

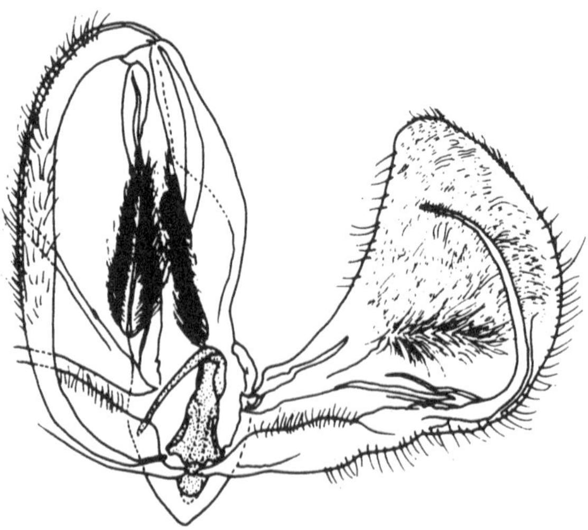

Fig. 44. *Pristoceraea eriopis* (HERRICH–SCHÄFFER).

ity of the areole; 10 from the upper angle thereof. In hind wings, vein 2 from very near the cell-angle; 3, 4 and 6, 7 from the cell-angles; discocellulars faintly incurved; 5 from middle of discocellulars.

Male genitalia: Uncus very elongate, one half again as long as the costa, narrow, with a small terminal hook and dense pilosity. Tubus analis about four-fifths of the uncus in length, very narrow, but broadening towards the extremity which is slightly excised. Valva narrow at base, then broadening to form a semi-circle; apex forming a nearly right angle; fold of the sacculus rather narrow, reaching to the middle of the valva; harpe very long and slender, strongly curved, with the extremity nearly reaching the costa. Aedeagus very small, slender, straight, with a slender, nearly filiform, curved and pointed terminal process; total length, with process, only three-quarters of the costa; fultura inferior broader proximally, narrow subterminally, with a lance-shaped extremity. Saccus hardly differentiated, rounded.

Distribution: Madagascar.

Pristoceraea eriopis (HERRICH-SCHÄFFER)

1853 *Agarista eriopis* HERRICH-SCHÄFFER, Außereurop. Schmetterl.: fig. 31. – Terra typica: Madagascar (Type, ♀, in British Museum).
1892 *Anaphela eriopis* – KIRBY, Syn. Cat. Lepidopt. Het.:25.
1895 *Pristoceraea eriopis* – KARSCH, Entomol. Nachr.Bl., Wien, **21**:350.
1896 *Rothia eriopis* aberr. *carminata* ROTHSCHILD, Novit. Zool., **3**:45. – Terra typica: Morondava (Type, ♂, in British Museum).
1901 *Ophthalmis eriopis* – HAMPSON, Cat. Lepidopt. Phal. Brit. Mus., **3**:646.
1913 *Rothia eriopis* – JORDAN in SEITZ, Großschmetterl. Erde, **15**:11; pl. 3d.
1974 *Eriopana eriopis* – KIRIAKOFF & VIETTE, Faune Madagascar, Lepidopt. Agaristidae, **39**: 82; fig. 66–68.

Diagnosis: Head, thorax above and fore wings black; palpi banded with white; sides of the frons, throat and spots on vertex, collar and tegulae white; pectus and legs orange; legs slightly streaked with black; abdomen orange, with dorsal and lateral series of black spots; the two last urites black in male; extremity of the anal tuft yellow. Fore wings with yellowish white spots at the base below the costa and in the base of the cell; an oblique bar across the extremity of cell; large spots beyond the cell, from interspace V to vein 11, and between veins 2 and 5; cilia white at the apex. Hind wings orange or carmine-red (aberr. *carminata*); base black; a terminal black band, broader apically and extending to the middle of the costa; a whitish spot beyond the upper cell-angle, rather indistinct in the typical form. Length of fore wing 20–22 mm.

Distribution: Madagascar.

Genus **Schausiana** KIRIAKOFF & VIETTE

(fig. 45)

1974 *Schausiana* KIRIAKOFF & VIETTE, Faune Madagascar, Lepidopt. Agaristidae, **39**: 97; fig. 80–81. – Species typica: *Eusemia obrysos* MABILLE (by original designation).

Diagnosis: Proboscis fully developed; palpi obliquely ascending; first and second joints with very dense pilosity; third very short, with short pilosity; frons

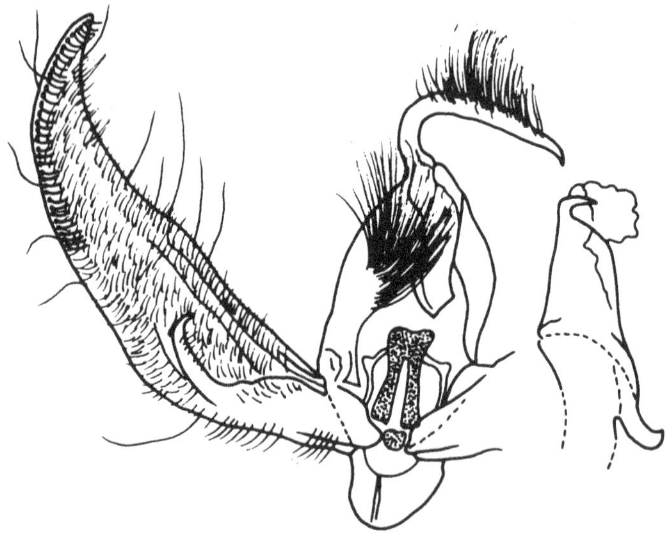

Fig. 45. *Schausiana obrysos* (MABILLE). Prep. P. Viette 5035. Paris Museum

of male very narrow so that the eyes are nearly conjoined; frontal prominence truncate conical, with terminal rim raised; antennae of male very slightly dilated towards the extremity, filiform in female; fore tarsi of male very elongate, with the basal joint stout, the remainder slender, the second joint being the longest, and the ventrolateral side bearing a series of very fine setae; fore wings in female shorter, with the termen less oblique and more rounded; nervation: in fore wings, vein 2 from four-fifths of cell; 3, 4 and 5 much approximated; discocellulars oblique, angled inwards; areole very narrow; 6, the common stalk of 7 + (8 + 9) and 10 from the extremity of the areole. In hind wings, vein 2 from very near to the cell-angle; 3 and 4 separate; 5 from middle of discocellulars; 6 and 7 from the upper cell-angle.

Male genitalia: Uncus rather short, about half the costa in length, slender, compressed, strongly arched at the base, then more or less straight, with a minute terminal hook, and very long and dense dorsal pilosity. Tubus analis shorter than the uncus, rather broad. Valva banana-shaped, tapering terminally to a point; last third of the sacculus bearing a marginal row of short setae; fold of sacculus short but broad; harpe short, placed rather proximally, ending in a small hook; fultura inferior narrow, forked for three-quarters of its length. Saccus semi-elliptical.

Distribution: Madagascar.

Schausiana obrysos (MABILLE)

1878 *Eusemia obrysos* MABILLE, Bull. Soc. zool. France, Paris, **3**:89. – Terra typica: Madagascar ex parte australe (Type, ♀, in Paris Museum).
1892 *Aegocera obrysos* – KIRBY, Syn. Cat. Lepidopt. Het.:37.
1913 *Schausia obrysos* – JORDAN in SEITZ, Großschmetterl. Erde, **15**:16; pl. 4b.
1974 *Schausiana obrysos* – KIRIAKOFF & VIETTE, Faune Madagascar, Lepidopt. Agaristidae **99**; fig. 82–83.

D i a g n o s i s : Head and collar yellow spotted with black; vertex orange in female; palpi light yellow; bases of the second and the third joints black; tegulae and thorax above mixed yellow and black; base of tegulae blackish; underside of thorax and legs pale yellowish orange; fore and mid-tarsi black; hind tarsi greyish brown, ringed with yellowish; abdomen orange with black basal tufts; anal tuft paler, with some black basally. Fore wings dark chocolate brown, sprinkled with silvery white, more densely on the dorsal half of the basal area; middle area much less sprinkled; a piriform white spot at one-third of the wing, broadening below the cell, and reaching vein 1; another spot, more elongate, at two-thirds of the wing, from costa to vein 3, with a distinct notch in interspace V; above tornus, a line without sprinkling, meeting the weakly sprinkeld middle area distally of the postmedian white spot; a terminal series of small spots or lunules without sprinkling; cilia pale grey with a dark medial line ant tip. Underside blackish brown, without sprinkling, and with the base orange. Hind wings bright orange in basal half, and black distally; the black terminal band about 8 mm wide costally, narrowing towards the anal angle where it is but 1 $^1/_2$–2 mm wide; a tooth on vein 2; cilia grey with a dark medial line and white spots. Underside similar, but with a white spot on the disco-cellulars. Length of forewing 18–21 mm.

D i s t r i b u t i o n : Madagascar.

Genus **Nesaegocera** KIRIAKOFF & VIETTE

(fig. 46)

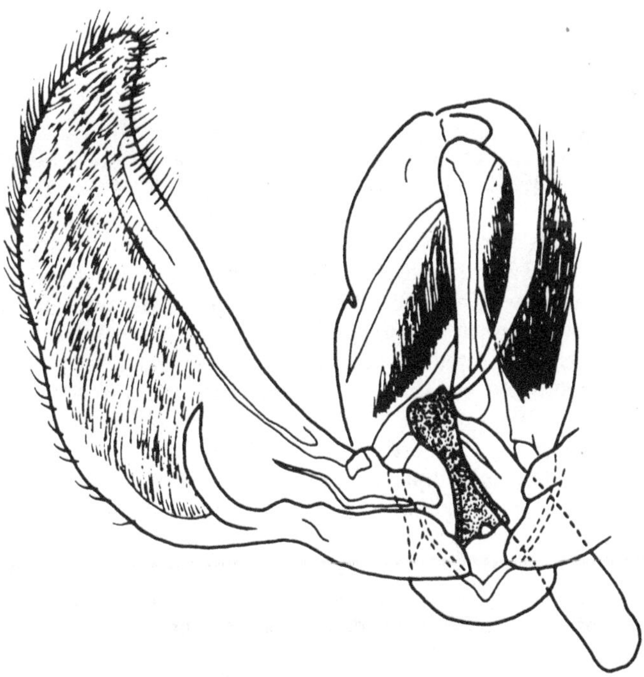

Fig. 46. *Nesaegocera comorana* (VIETTE).

1974 *Nesaegocera* KIRIAKOFF & VIETTE, Faune Madagascar, Lepidopt. Agaristidae, **39**: 102; fig. 84–85. – Species typica: *Aegocera comorana* JORDAN (by original designation).

Diagnosis: Proboscis fully developed; palpi obliquely upturned, nearly reaching the upper edge of the genae; first and second joints densely hairy; third joint slightly shorter than the second, slender, naked, slightly expanded terminally; frons with a rounded prominence; antennae hardly expanded towards the extremity; legs with the pilosity rather dense but short. Female with the fore wings distinctly more elongate than the male. Nervation: vein 2 from about five-sixths of the cell; 3, 4 and 5 well separated, equidistant; discocellulars very short; 6 from just below the upper cell-angle; areole very narrow; 7 and the stalk of (8 + 9) and 10 from ghe extremity of the areole. In hind wings, vein 2 from very near the angle of cell; 3, 4 and 6, 7 from the angles; 5 from middle of the discocellulars which are angled inwards.

Male genitalia: Uncus about two-thirds of the costa in length, narrow and compressed, arched, more strongly at the base. Tubus analis slightly shorter than the uncus, narrow, faintly expanded terminally. Valva arched, broad in the middle portion, with the apex more strongly upcurved and more acute; costa regularly incurved; sacculus strongly arched distally of the insertion of the harpe; the latter short, arched. Aedeagus slightly shorter than the uncus, rather robust, nearly straight, unarmed; fultura inferior narrow, especially in the middle portion; proximal margin with the angles somewhat produced, and with a median excision. Saccus very short, rounded.

Distribution: Comoro Archipelago.

Nesaegocera comorana (JORDAN)

1926 *Aegocera comorana* JORDAN, Novit. zool., **33**:374. – Terra typica: Grand Comoro (Type, ♀, in British Museum).
1974 *Nesaegocera comorana* – KIRIAKOFF & VIETTE, Faune Madagascar, fasc. 39, Lepidopt. Agaristidae **39**: 104; fig. 86–87.

Diagnosis: Male. Palpi black; base of the last joint and pilosity of the two first joints orange ochre; head orange ochre with a brown median line; thorax above chocolate brown with an ochraceous median line; collar ochraceous, spotted with chocolate-brown; underside and legs orange ochre; fore legs streaked with brown above; mid tibiae spotted with black; mid tarsi ringed ochraceous and black; abdomen orange ochre with squarish black dorsal spots; anal tuft somewhat mixed with chocolate brown. Fore wings dark chocolate brown, rather densely sprinkled with orange ochre, less so on the proximal tow-thirds of costa and on dorsal area; inner line represented by an oblique brown line below vein 1 preceded by a subbasal streak; cell-mark small, rounded; discocellular mark squarish, both marks sprinkled with silvery blue in median portion, and edged with chocolate brown; outer line from slightly beyond two-thirds of the costa, very fine, broadly excurved to vein 2, then broadening, becoming nearly vertical and not reaching beyond vein 1; distal area somewhat darker; terminal line chocolate brown, very fine; cilia greyer. Underside duller, with the typical marks brown and a pale chocolate brown marginal band broadening towards the apex. Hind wings orange ochre with a pale brown discocellular mark and a brown terminal band, 3 mm wide in the costal region, darker chocolate brown towards the anal angle; marginal line dark chocolate brown with pale ochraceous internervural lines. Underside with the terminal band paler. Female: head and thorax above yellowish white; abdomen lemon yellow. Fore wings pale lemon yellow, densely sprinkled with vandyck brown, paler in the distal third; discal area white, that colour beginning at the base and occupying the dorsal half of the cell, the proximal half of interspace I and the bases of interspaces II–VI; cell-mark reduced to a point; discocellular mark as in male. Hind wings light lemon yellow with the terminal band only present as a slight brown sprinkling. Length of fore wing 17–21 mm.

Distribution: Comoro Archipelago (Grand Comoro).

Genus **Arctiopais** JORDAN
(fig. 47)

1896 *Arctiopais* JORDAN, Novit. Zool., 3:46. – Species typica: *Hypsa ambusta* MABILLE (by monotypy).

Diagnosis: Proboscis fully developed; first and second joints of palpi upturned, fringed with long hair in front; third joint very long, smoothly scaled, porrect; frons narrowed behind in male; frontal prominence less than one-third of frons in breadth, truncate conical, with a raised terminal rim; antennae distinctly dilated towards the extremity in male, less so in female; tibiae with long upstanding hair; in male also mid- and hind tarsi and spurs hairy; hind tibiae with two pair of spurs, the distal pair very long. Nervation: in fore wing, vein 2 from seven-tenths of

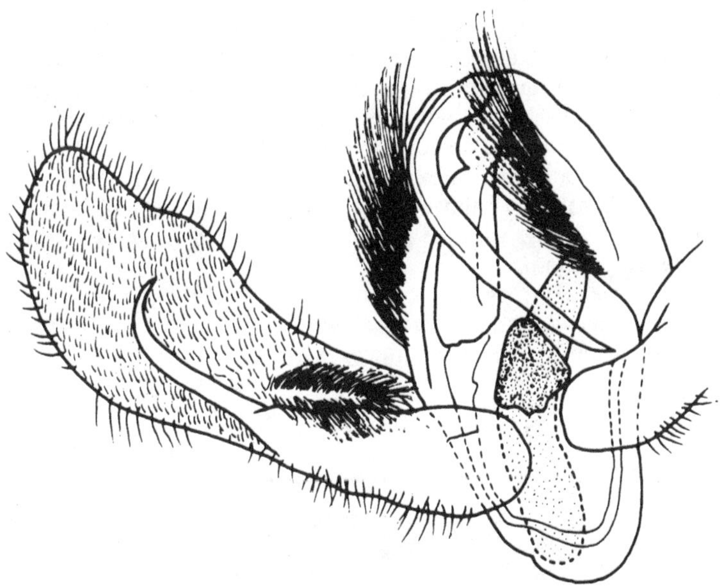

Fig. 47. *Arctiopais ambusta* (MABILLE).

cell; 3, 4 and 5 approximated, equidistant; discocellulars faintly incurved; areole narrow; 6 from the lower cell-angle; 7 and stalk of 8 + 9 from the lower angle of areole; 10 from the upper angle thereof. In hind wings, vein 2 from five-sevenths of the cell; 3, 4 and 6, 7 from the cell-angles; discocellulars faintly incurved; 5 obsolescent, from middle of the discocellulars.

Male genitalia: Uncus slightly shorter than the costa, rather slender, compressed, strongly arched in the basal quarter, then nearly straight. Tubus analis shorter than the uncus, rather broad. Vinculum with the fold narrow. Valva elongate, nearly straight, broadening distally; apex rounded; costa slightly jutting out at two-thirds; fold of sacculus very broad basally, narrowing distally and not reaching the middle of the valva; harpe long, arched in the terminal third. Aedeagus about as long as three-quarters of the costa, robust, slightly angled at middle; fultura inferior triangular, with the proximal margin produced and slightly notched at middle, and the distal margin rounded. Saccus semi-elliptical.

Distribution: Madagascar.

Arctiopais ambusta (MABILLE)

1881 *Hypsa ambusta* MABILLE, C. R. Soc. entomol. Belgique, Bruxelles, **25**:55. – Terra typica: Morondava (Type, ♀, in British Museum).
1892 *Pseudhypsa ambusta* – KIRBY, Syn. Cat. Lepidopt. Het.:384.
1896 *Arctiopais ambusta* – JORDAN, Novit. Zool., **3**:47.
1913 *Arctiopais ambusta* f. *melanura* JORDAN in SEITZ, Großschmetterl. Erde, **15**:12. – Terra typica: Morondava (Type, ♀, in British Museum).
1926 *Arctiopais ambusta* ssp. *celis* JORDAN, Novit. Zool., **33**:374. – Terra typica: Diego Suarez (Type, ♂, in British Museum).

Diagnosis: Body and wings orange; last joint of the palpi, antennae and tarsi black; abdomen with a black dorsal patch on the terminal segment (♂), or the terminal segment black (♀). Fore wings with a few silvery scales below the basal part of the costa; an irregular oblique curved interrupted antemedian line with spots beyond it in the cell and above vein 1; an indistinct blackish postmedian line with some silvery beyond it, oblique from the costa to vein 4, then excurved to the submedian fold and incurved to the dorsum; area beyond it to termen suffused with deep rusty red in female, with a pale mark above the tornus; costal and terminal lines black; cilia silvery grey. Hind wings with the terminal two-thirds black; cilia silvery grey. Length of fore wing 20 mm.

The form *melanura* JORDAN differs in having the tip of the abdomen black in both sexes. The form *celis* JORDAN has the terminal area in fore wings, the black terminal band narrower, 7 mm wide at the apex and only 1 mm towards the anal angle.

Distribution: Madagascar.

Genus Arrothia JORDAN

(fig. 48)

1896 *Arrothia* JORDAN, Novit. Zool., **3**:46. – Species typica: *Arrothia bicolor* ROTHSCHILD & JORDAN (by original designation).

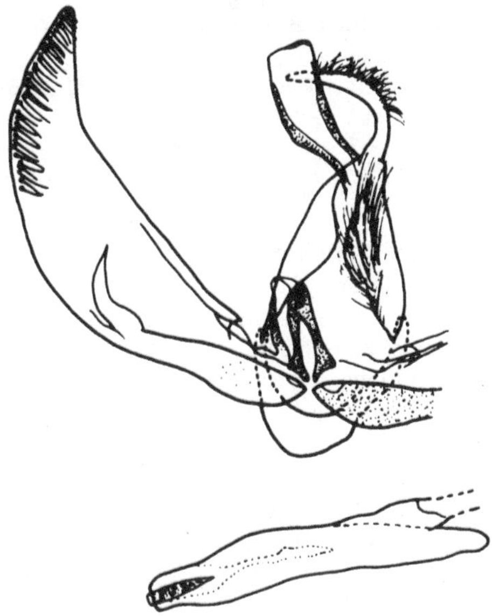

Fig. 48. *Arrothia bicolor* (ROTHSCHILD & JORDAN). B. M. Noct. 8535

Diagnosis: Proboscis fully developed; palpi upturned; second joint with rather sparse hair in front; third long, smooth, porrect; frons with a long horny spine truncate at the extremity; antennae distinctly dilated towards the extremity; legs fringed with long hair. Nervation: in fore wings, vein 2 from seven-tenths of cell; 3, 4, 5 much approximated; discocellulars faintly incurved; 6 from the upper cell-angle; areole short; 7 and the long stalk of 8 + 9 from the lower angle of the areole; 10 from the upper angle thereof. In hind wings, vein 2 from three-quarters of cell; 3, 4 and 6, 7 stalked; discocellulars very faintly incurved; 5 weak, from middle of the discocellulars.

Male genitalia: Uncus typically shorter than half costa, sickle-shaped. Tubus analis as long as the uncus, widening distally. Vincula broader in the median portion. Valva shaped like a banana, but with the apical portion more produced, and pointed; apex and distal third of the sacculus with a marginal series of spinulae; fold of the sacculus very broad relatively to the breadth of the valva at the base, not quite reaching one third of the valva; harpe short, arched. Aedeagus slightly shorter than the costa, rather robust, nearly straight; vesica bearing a few cornuti; fultura inferior shaped like a V placed upside down, with the branches expanding at the extremities. Saccus short, rounded. The male genitalia of *A. gueneeiana* Viette offer a number of differences. Uncus less arched; valva broader. upturned, narrowing towards the apex; harpe placed vertically and much more robust; aedeagus rather slender; fultura inferior narrow, with the terminal margin produced and bifid. Some students might be tempted to place *A. gueneeiana* in a genus of its own.

Key to the species

1 Fore wings black at the base and in the distal half . . . . **A. bicolor**, p. 148
1' Fore wings have only a fine apical line and cilia black . . . . **A. gueneeiana**, p. 148

Arrothia bicolor ROTHSCHILD & JORDAN

1896 *Arrothia bicolor* ROTHSCHILD & JORDAN, Novit. Zool., **3**:46; pl. 14, fig. 20. – Terra typica: Morondava (Type, ♀, in British Museum).
1926 *Arrothia bicolor bicolor* – JORDAN, Novit. Zool., **33**:373.
1926 *Arrothia bicolor melanobasis* JORDAN, Novit. Zool., **33**:373. – Terra typica: Lake Aloatra (Type, ♂, in British Museum).

Diagnosis: Head black; thorax above buffish orange; underside and legs black; abdomen orange, with the last two urites and the underside black. Fore wings with the base black with a few orange scales; proximal half orange; distal half black, with the proximal margin running from two-fifths of the costa to five-sevenths of the dorsum. Hind wings orange basally, with a few blackish scales; terminal area black, slightly wider than the orange basal portion. Length of fore wing 19 mm.

The form *melanobasis* JORDAN has the orange area of the fore wings reduced to an antemedial band about 2,5 mm wide, not quite reaching the cell-end, and much mixed with black at the base; in hind wings, orange area likewise reduced, and the base black for about 3 mm.

Distribution: Madagascar.

Arrothia gueneeiana VIETTE

1954 *Arrothia gueneeianum* VIETTE, Mém. Inst. scient. Madagascar, (E) **5**:71; textfigs. – Terra typica: Ankarafantsiaka (Type, ♂, in Paris Museum).

Diagnosis: Head, antennae and palpi black; vertex, and thorax above ochreous; underside of the abdomen and legs black; abdomen ochreous proximally; last urites and underside black. Fore wings uniform orange red; a fine black marginal line at the apex; cilia black. Hind wings also uniform orange red, with a black terminal band about 3 mm wide. Length of fore wing 16–17 mm.

Distribution: Madagascar (northwestern part).

Genus **Antigodasa** KIRIAKOFF & VIETTE

(fig. 49)

1974 *Antigodasa* KIRIAKOFF & VIETTE, Faune Madagascar, Lepidopt. Agaristidae, **39**: 94; fig. 78. – Species typica: *Godasa rufodiscalis* ROTHSCHILD (by original designation).

Diagnosis: Proboscis fully developed; palpi reaching beyond the upper margin of the eyes, oblique, with the pilosity moderate; last joint hanging; antennae nearly filiform. Nervation: in fore wings, vein 2 from four-fifths of the cell; 3, 4, 5 separate, equidistant; areole small; 6 from the lower cell-angle; 7, 8 + 9 and 10 from the extremity of the areole. In hind wings, vein 2 from three-quarters of the cell; 3, 4 and 5, 7 from the cell-angles; 5 from the middle of the discocellulars, not weaker than the remaining veins.

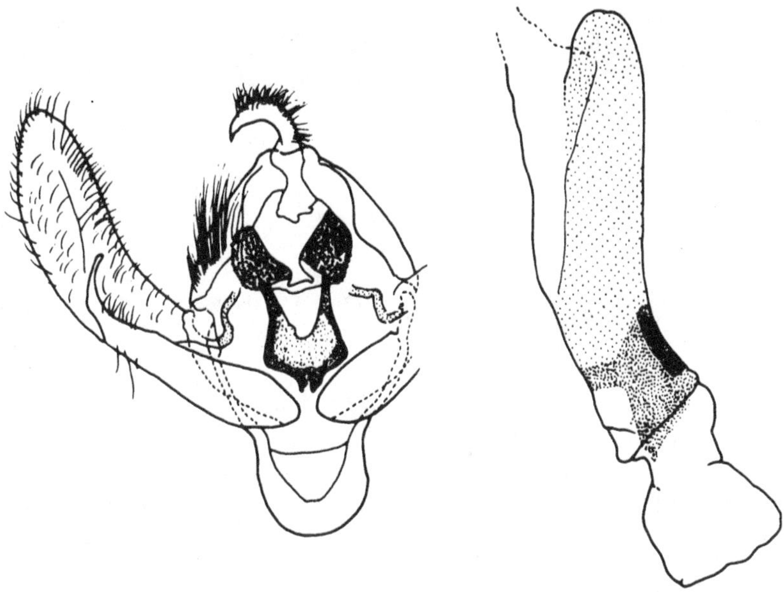

Fig. 49. *Antigodasa rufodiscalis* (ROTHSCHILD). B. M. Noct. 8289. Type

Male genitalia: Uncus small and weak, hooked. Tubus analis short, widening distally. Vinculum with the fold narrow, bearing the usual bunch of long hair. Valva elongate and narrow, slightly upturned, with the distal portion tongue-shaped; fold of the sacculus broad but short; harpe short, arched. Aedeagus relatively enormous, much longer than the valva, very robust, arched in distal third; fultura inferior lyre-shaped, broad proximally, with the arms tapering to a point; fultura superior consisting of two large lateral lobes covered with minute spinulae, and connected by means of a median membrane. Saccus elongate semi-elliptical.

Distribution: Madagascar.

Antigodasa rufodiscalis KIRIAKOFF & VIETTE

1896 *Godasa rufodiscalis* ROTHSCHILD, Novit. Zool., **3**:47; pl. 14, fig. 23. – Terra typica: Madagascar (Type, ♂, in British Museum).
1901 *Tuerta (Misa) rufodiscalis* – HAMPSON, Cat. Lepidopt. Phal. Brit. Mus., **3**:626.
1912 *Tuerta rufodiscalis* – STRAND, Lepidopt. Cat., **5**:29.
1974 *Antigodasa rufodiscalis* – KIRIAKOFF & VIETTE, Faune Madagascar., fasc. 39, Lepidopt. Agaristidae, **39**: 96; fig. 79.

Diagnosis: Head and thorax above blackish brown, slightly sprinkled with whitish, and marked with white; thorax below and legs blackish brown; tarsi ringed with white; abdomen orange; basal and anal tufts black; a lateral and median series of black spots. Fore wings dark brown, with a chocolate tinge; scattered silvery blue scales at the base, towards middle of the cell, below the cell, on the discocellulars, and postmedially, where they form two wavy lines, one postmedial, the other subterminal; a terminal series of white dots, the one above the dorsum much the largest. Hind wings with the base and the costal area blackish brown; discal area reddish orange; a black terminal band, broad apically, narrowing towards the anal angle; discocellulars with a black streak. Length of fore wing 21 mm.

Distribution: Madagascar.

Genus **Ancarista** JORDAN
(fig. 50)

1921 *Ancarista* JORDAN, Novit. Zool., **28**:73. – Species typica: *Ovios laminifera* SAALMÜLLER (by original designation).

Diagnosis: Proboscis fully developed; palpi short; first and second joints with the pilosity forming a triangle; third porrect, short, smooth, slightly dilated terminally; antennae filiform; frontal prominence truncate conical; legs with the pilosity short; abdomen with large basal crest. Fore wings in male narrower, and with the termen more oblique than in female. Nervation: Fore wings with vein 2 from four-fifths of the cell; 3, 4, 5 separated; 6 from the upper cell-angle; discocellulars incurved; areole elongate and narrow; 7, the short stalk of 8 + 9, and 10 from the extremity of the areole. Hind wings with vein 2 given off very near the cell-angle; 3, 4 and 6, 7 from the cell-angles; discocellulars nearly straight; 5 obsolescent, from middle of the discocellulars.

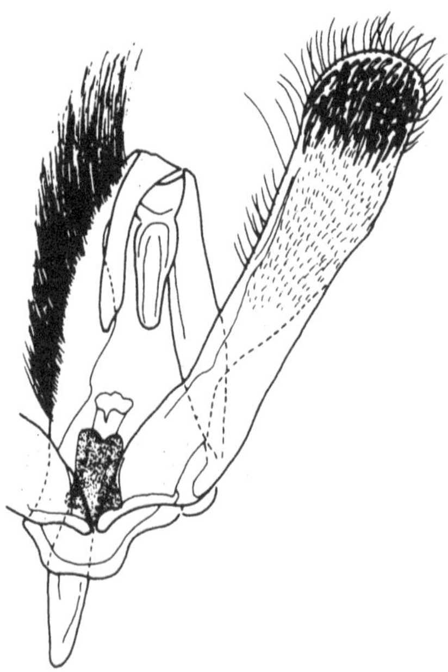

Fig. 50. *Ancarista laminifera* (SAALMÜLLER).

Male genitalia: Uncus short, less than half of valva in length, narrow, faintly arched. Tubus analis still shorter, narrowing distally. Vinculum narrow, with a fringe of partly very long setae. Valva very elongate and narrow; costa and sacculus nearly straight and parallel; apex broadly rounded; terminal portion covered with stout spines; harpe absent. Aedeagus about half of valva in length, rather slender, especially distally, straight; fultura inferior rounded, with the proximal margin produced at middle, and the distal margin bilobate.

Distribution: Comoro Archipelago and Madagascar.

Ancarista laminifera (SAALMÜLLER)

1878 *Ovios laminifera* SAALMÜLLER, Ber. Senckenb. naturf. Ges., **1878**:91. – Terra typica: Sambirano, Nosi Bé (Type, ♀, in Senckenberg Museum).
1884 *Euschirropterus laminifera* – SAALMÜLLER, Lepidopt. Madagascar:140; fig. 133.
1910 *Paratuerta laminifera* – HAMPSON, Cat. Lepidopt. Phal. Brit. Mus. **9**:73.
1921 *Ancarista laminifera* – JORDAN, Novit. Zool., **28**:73.

Diagnosis: Sexes rather different. Male: Head white; first and second palpal joints white, brown laterally; third joint chocolate-brown, with the underside and the extremity white; thorax above white; collar white whith lateral chocolate-brown marks; tegulae chocolate-brown, slightly sprinkled with white; underside and legs white; fore tibiae chocolate-brown; tarsi light rufous brown; basal abdominal crest chocolate-brown, sprinkled with white; abdomen ochreous orange. Fore wings umberbrown in the middle area of the costal region and in the dorsal half; rest of the costal region and a broad median stripe, white, separated from the brown dorsal area by a fine silvery edge, followed by a blackish chocolate streaklet from the base to the costa before the apex; the white median area tinged with brown distally; angles of the discocellulars each with a brown dot; a narrow whitish basal line; a subbasal blackish chocolate line, fine, arched and rejoining a similar but more distal line; both finely edged with white; dorsum narrowly blackish chocolate at the base; along dorsum, an elongate whitish elliptical spot; terminal area becoming light grey, with a whitish undulate subterminal line; terminal line broken, blackish chocolate; costa spotted with blackish chocolate towards apex; cilia light grey, spotted with brown at the extremity. Hind wings bright ochraceous, becoming whitish in the costal area; terminal band brown, 1–2 mm wide, with a median white line, and with the proximal half broken into spots; apex whitish; cilia white, with a brown median line. Female: palpi white, sprinkled with chocolate brown; third joint as in male; frons and vertex chocolate-brown; collar and thorax above chocolate brown laterally, white sprinkled with chocolate-brown medially; tegulae with an indistinct whitish median strek; thorax below and legs orange yellow; fore legs streaked with chocolate-brown above; basal abdominal crest chocolate-brown; rest of the abdomen yellow orange. Fore wings wood-brown, slightly suffused with chocolate-brown; dorsal area darker, suffused with blackish; a silvery streak from the base to the outer line below the cell, toothed at the first quarter, and subterminally, with the result that the median portion of the streak is wider; cell-angles each with a dark dot; costal and dorsal areas with dark streaks; base of the wing blacker dorsally; outer line from a little before the apex, faintly arched, limiting the dark portion of the wing; terminal area becoming grey, and marked as in male. Hind wings orange yellow, with a dark brown terminal band, 2 mm wide, not reaching the anal angle; before the latter, a streak 2 mm long towards the base; cilia tipped with whitish, especially in the costal area. Length of fore wing 18–19 mm.

Ancarista laminifera laminifera (SAALMÜLLER)

1966 *Ancarista laminifera laminifera* – VIETTE, Lambillionea, **65**:89.

Diagnosis: As above.

Distribution: Madagascar.

Ancarista laminifera comorana VIETTE

1966 *Ancarista laminifera comorana* VIETTE, Lambillionea, **65**:89. – Terra typica: Grand Comoro (Type, ♀, in Paris Museum).

Diagnosis: Fore wings greyer, especially distally; the silvery streak has no tooth on its lower edge. Hind wings less bright yellow, and the terminal black band slightly wider.

Distribution: Comoro Archipelago.

Genus **Musurgina** JORDAN
(fig. 51)

1921 *Musurgina* JORDAN, Novit. Zool., **28**:71. – Species typica: *Musurgina laeta* JORDAN (by original designation).

Diagnosis: Proboscis fully developed; palpi clothed beneath with rough hair; last joint porrect; frons with a long conical process, flattened above and below, truncate with a terminal rim; antennae dilated towards the extremity, and with a terminal hook; prothorax with a scaly crest; hind tibiae with two pair spurs, the pro-

Fig. 51. *Musurgina laeta* JORDAN.

ximal pair being short, and with a naked dorsal stripe of transverse ridges; base of the abdomen with a forked crest. Nervation: in fore wings, vein 2 given off very near the angle of cell; 3, 4, 5 separate; 6 from the upper cell-angle; areole small; 7 from the extremity of the areole, short stalked with 8 + 9; 10 from the upper side of the areole. In hind wings, vein 2 from very near the angle of cell; 3, 4 and 6, 7 from the cell-angles; 5 from middle of the discocellulars.

Male genitalia: Uncus distinctly shorter than the costa, very slender, and strongly arched. Tubus analis short, rather narrow. Fold of vinculum broad, with dense pilosity. Valva banana-shaped; apex subacute; sacculus with a dense marginal series of stiff setae, placed subterminally and terminally; fold of sacculus broad proximally, narrowing distally, rather short; harpe short and slender, hooked. Aedeagus hardly longer than half of costa, rather slender, nearly straight, bearing at middle a patch of minute cornuti; fultura inferior shaped like an inverted triangle. Saccus semi-elliptical.

Distribution: Madagascar.

Musurgina laeta JORDAN

1921 *Musurgina laeta* JORDAN, Novit. Zool., **28**:71. – Terra typica: Diego Suarez (Type, ♂, in British Museum).

Diagnosis: Male: Head and thorax above grey russet; shoulders white; collar whitish, edged with brownish grey; tegulae grey brown, edged with white, and with black extremities; underside and legs greyish; tarsi grey russet, with the basal crest ochreous, tipped laterally with black. Fore wings brownish ochreous grey, more ochreous in dorsal and subterminal areas, more brownish in the costal region; an elongate hyaline spot in cell, another below cell; a dark brown diffuse band along the dorsal cell-margin; interspaces I and II with a terminal rusty dot, edged proximally with white, and placed in a semi-circular grey brown terminal spot. Hind wings yellow, becoming more orange towards the anal margin; a blackish brown terminal band 2 mm wide, with the outline somewhat irregular; cilia yellowish, spotted with rusty in interspace I and towards the anal angle. Female: rather different. Fore wings much less produced at the apex; costal margin not incurved subapically; termen much less oblique. Fore wings greyish umber brown slightly suffused with orange, except in the costal area, and on a terminal band 2mm wide, which are very dark grey; a broad silvery white slightly hyaline stripe from the base to middle of the wing, then curving upwards and reaching the costa 2 mm before the apex; termen with a series of rufous, white-edged dots; the white edgings more distinct in interspaces I–III, where they fuse to form a band. Length of fore wing 19–20 mm.

Distribution: Madagascar.

Genus **Pemphigostola** STRAND

(fig. 52)

1909 *Pemphigostola* STRAND, Dtsch. entomol. Z., **1909**:665. – Species typica: *Pemphigostola synemonistis* STRAND (by original designation).

1926 *Pemphigostala* STRAND (ex errore pro *Pemphigostola*), in SEITZ, Großschmetterl. Erde, **15**:16.

D i a g n o s i s : Proboscis full developed; palpi short, oblique; first and second joints with long frontal pilosity; third joint short, smooth, pointed; frons with truncate conical prominence; antennae of male club-shaped terminally; antennae of female filiform; legs with the pilosity moderate. Fore wings in male triangular, strongly inflated and outcurved costally owing to the presence of a basal vesicle; termen straight, very oblique; tornus broadly rounded. Nervation: in fore wings, vein 2 from two-thirds of cell; 3 and the stalk of 4–5 from the lower cell-angle; areole exceedingly narrow; 6–10 from the extremity of the areole. Female without a stridulating organ; cell very long and narrow; vein 2 from four-fifths of the cell; 3, 4, 5 much approximated; discocellulars angled inwards; areole very small, elongate; 6, the common stalk of 7 and 8 + 9, and 10 from the extremity of the areole. In hind wings (male and female), vein 2 from four-fifths of the cell; 3–4 and 6–7 very short stalked, from the cell-angles; 5 from the middle of the discocellulars, not weaker than the other veins.

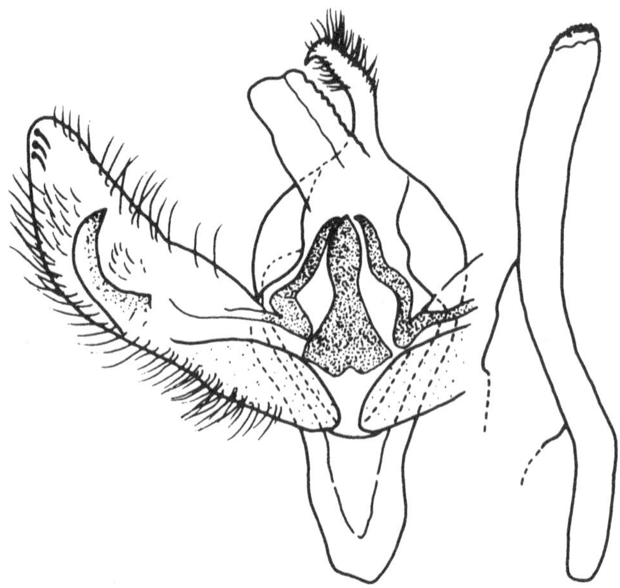

Fig. 52. *Pemphigostola synemonistis* STRAND.

Male genitalia: Uncus relatively short and narrow, arched, rounded at the extremity. Tubus analis short and very broad. Fold of vinculum without the usual pilosity. Valva elongate, upcurved, rather broad; costa faintly convex for about two-fifths, then nearly straight; apex rounded, with a few stiff spinulae; saccus more or less regularly convex, with a faint incurving beyond the middle; fold of sacculus broad and short; harpe robust, arched, placed in the distal half of the valva. Aedeagus relatively very long, longer than the valva, moderately robust, faintly angled at one-fourth, then somewhat arched; vesica bearing cornuti; fultura inferior broad proximally; proximal margin excised at middle, with the angles produced; distal margin rounded. Saccus elongate semi-elliptical.

The genus *Pemphigostola* was placed by its describer in the family Castniidae, subfamily Pemphigostolinae; HAMPSON (Novit. Zool., **25**:377, 1918) placed it in the family Apoprogenidae; HERING (Biol. Schmetterl., in Biol. Studienbücher, **3**, pl. 4, fig. 4, 1926) referred *Pemphigostola* to the family Noctuidae. JORDAN (Novit. Zool., **28**:73, 1921) was the first to place the genus with the Agaristidae, albeit tentatively; HERING (in BLÜTHGEN & HERING, Entomol. Zeitschr., **42** (14): 176, 1928) confirmed the opinion of JORDAN, which was definitively established by VIETTE (Bull. Soc. entomol. France. **60** (1955): 176, fig. 1–5, 1956), who was the first to study the male genitalia of *Pemphigostola*.

Distribution: Madagascar.

Pemphigostola synemonistis STRAND

1909 *Pemphigostola synemonistis* STRAND, Dtsch. entomol. Z., **1909**:665. – Terra typica: Madagascar (Type, ♂, in Berlin Museum).

Diagnosis: Sexes different. Male: Palpi and periocular region vandyck brown; vertex light grey; frontal prominence dark brown; collar vandyck brown, edged with whitish; tegulae and thorax above dark brown, the former with a diffuse whitish spot; metathorax with a white median stripe; thorax below and legs grey brown; abdomen grey brown with a whitish dorsal stripe; segmental edges and rather obsolete sublateral and lateral series of spots, whitish; anal tuft blackish brown. Fore wings rather light grey brown; base of the costa and the proximal two-thirds of the dorsal region vandyck brown, sprinkled with orange; at the first third of the dorsal region, a fine brown line, ending in a broad dorsal lunule; median cell and a streak below it, hyaline, with yellowish brown veins; interspace I with a basal blackish brown dot; the orange brown colour limited distally by a short blackish brown line, followed by a whitish one, and by a blackish brown spot; towards the middle of the wing, an elongate and narrow white subcostal spot, followed by a vandyck brown spot, sprinkled with orange, and by a subtriangular white costal spot reaching vein 5 where it is truncate, and bearing a slight grey line; outer line fine, blackish, edged with white proximally, limiting distally the white spot, broadly incurved to interspace I, where it curves down towards the tornus; a subapical vandyck brown spot, mixed with some orange; termen whiter, with a terminal series of small blackish streaklets; cilia with the base and spots brown. Hind wings orange in the proximal two-thirds, with an umber brown band from the middle of the costa, through the cell, broadening at the bases of interspaces II and III, and diffuse between vein 2 and the anal margin; median cell tinged with umber brown proximally; its distal part greyer and merging into the brown band, dividing the orange area into two portions; a terminal series of blackish brown dots; cilia whitish, tipped with brown. Female: body as in male. Fore wings dark sepia grey; costa whitish, sprinkled with brown; basal area suffused with orange, but the extreme base dark brown; inner line fine, white, very oblique from near the base to the dorsum 3 mm from the tornus; a broad angulate white spot, finely speckled on the margins with brown, from the costa to vein 4, with an oblique black line on it; outer line fine, white, faintly incurved, limiting distally the costal spot, and reaching the tornus at 2 mm from the inner line; area between the lines somewhat darker; termen paler towards the tornus; an apical whitish triangular spot; between this and the proximal white spot, a chestnut brown suffusion; cilia whitish, tipped with brown. Hind wings orange; marginal band dark grey brown, broader (about 5 mm) and paler costally and in the anal region; an indistinct grey brown spot across the cell; a fine subterminal line from about 2 mm from the anal angle, becoming diffuse and indistinct towards interspace IV; before the anal angle, a short whitish line; between both lines, a rounded spot of whitish sprinkling; cilia as in fore wing. Length of fore wing 18–21 mm.

Distribution: Madagascar.

Addendum

Since the manuscript has been sent to the editor, the following new species of African Agaristidae have been described:

Heraclia aurantiaca KIRIAKOFF, Rev. Zool. Afr., 1975, 89 (1):55, fig. 15.
Schausia dambuza KIRIAKOFF, ib., 78, fig. 37.
Sch. mantatisi KIRIAKOFF, ib., 79, fig. 38.
Heraclia nandi KIRIAKOFF, Bull. Ann. Soc. r. belge Ent., 1974, 110:146, fig. 1.
H. pampata KIRIAKOFF, ib., 147, fig. 2.
H. nobela KIRIAKOFF, ib., 149, fig. 3.
Schausia mkabi KIRIAKOFF, ib., 151, fig. 5.
Sch. langazana KIRIAKOFF, ib., 152, fig. 6.
Massaga noncoba KIRIAKOFF, ib., 154, fig. 7.
Heraclia jacksoni KIRIAKOFF, ib., 1975, 111:197, fig. 1.
Pseudagoma gen. nov. KIRIAKOFF, ib. 199.
Ps. pinheyi KIRIAKOFF, ib., 200, fig. 2.
Brephos sublaeta KIRIAKOFF, ib., 202.

The following have inadvertendly been omitted:

Ad p. 64.

Aegoceropsis tricolora (BETHUNE-BAKER)

1909 *Aegocera tricolora* BETHUNE-BAKER, Ann. Mag. nat. Hist., London, (8) 3:434. – Terra typica: Kambove (Type, ?, in British Museum).

Diagnosis: Head and thorax black, stripped white; pectus orange; femora and tibiae with orange hair; abdomen reddish orange with a dorsal black stripe; anal tuft and underside orange, the latter with a sublateral series of black spots. Fore wings black; a broad subdorsal stripe of red, from base to tornus; other markings white, viz., a subbasal subcostal dot; an antemedial wedge-shaped subcostal mark; an oblique band from media to submedian fold; an oblique postmedian band from below costa to below vein 3, constricted at middle; silvery blue scales forming a double discocellular line, and another, distally of the postmedian band; some white scales in the terminal area; cilia silvery blue at tips. Hind wings orange or red, with a black terminal band, wider in the median portion; cilia with white tips, becoming silvery blue below vein 4. Length of fore wing 15–22 mm.

Distribution: Zaire.

Aegoceropsis tricolora tricolora (BETHUNE-BAKER)

Diagnosis: Hind wings orange.

Distribution: Zaire: Bandundu, Shaba.

Aegoceropsis tricolora postrubra KIRIAKOFF

1974 *Aegoceropsis tricolora postrubra* KIRIAKOFF, Rev. Zool. Afric., Tervuren, 89(I):66. Terra typica: Zaire, Isiro (Type, ♂, in Museum Tervuren).

Diagnosis: Hind wings red.

Distribution: Zaire: Province Orientale, Equateur.

Ad p. 111.

Tuerta chrysochlora chrysochlora WALKER

Diagnosis: As above.

Distribution: Gaboon, Nigeria, Zaire.

Tuerta chrysochlora liturata (AURIVILLIUS)

1892 *Eudryas?* liturata AURIVILLIUS, Ent. Tidskr., 13:187. Terra typica: Gold Coast (Type, ?, in Stockholm Museum).

1892 *Metagarista aziyade* VUILLEMIN, Bull. Soc. ent. France, 61:140. Terra typica: Sierra Leone (Type in British Museum).

Diagnosis: Marginal band of fore wings broader and straight proximally.

Distribution: Gold Coast, Cameroons, Sierra Leone.

Alphabetic Register of Scientific Names

abacata, Xanthospilopteryx 28
abyssinica, Aegocera 67
–, Charilina amabilis 95
–, Mitrophrys 67
–, Mitrophrys magna 67
Acantuerta 4, 122
accra, Charilina amabilis 95
adulatrix, Eusemia 16
–, Heraclia butleri 16
–, Rothia butleri 16
–, Xanthospilopteryx 16
Aegoceropsis 1, 2, 3, 58
aemulatrix aemulatrix, Heraclia 32
– campala, Heraclia 32
– campala perpallida, Xanthospilopteryx 32
– campala, Xanthospilopteryx 32
–, Eusemia 31
–, Heraclia 31
–, Heraclia aemulatrix 32
– suda, Heraclia 32
– suda, Xanthospilopteryx 32
–, Xanthospilopteryx 31
aequipartita, Aegocera kibwezensis 67
Aethodes 3, 56
affinis, Aegocera 58
–, Aegocera fervida 59
–, Aegoceropsis 58
africana, Eusemia 19
– flava, Xanthospilopteryx 19
–, Heraclia 19
–, Xanthospilopteryx 19
Agoma 3, 70
agoma, Aegocera trimeni halans 72
–, Mitrophrys 72
agrius, Agarista 134
–, Anaphela 135
–, Eusemia 134
–, Rothia 134
aisha, Heraclia 24
–, Rothia 24
– transiens, Xanthospilopteryx 24
–, Xanthospilopteryx 24
alba, Leucovis 122
–, Ovios 122
–, Perimagia 122
–, Pristoceraea 122
albifascia, Aegocera tigrina 54
albigutta, Charitosemia 117
–, Mimeusemia 117
–, Pristoceraea 117
albiventrix, Aegocera trimeni trimeni 72
albocincta, Heraclia 17
Aletopus 1, 2, 10
alice, Xanthospilopteryx kirbyi 31
alienata, Noctua 95
allaudi, Rothia 130
alluaudi, Phalaenoides 130
–, Rothia 130
–, Rothia rhaeo 130
amabilis abyssinica, Charilina 95
– accra, Charilina 95
–, Aegocera 95
–, Charilina 95
–, Crameria 95
–, Noctua 95
– nyassica, Charilina 95
Amazela 44
ambusta, Arctiopais 146
– celis, Arctiopais 146
–, Hypsa 146
– melanura, Arctiopais 146
–, Pseudhypsa 146
Ancarista 125, 150
Andrhippuris 1, 2, 4
angolana, Chaetostephana rendalli 43
angustata, Charitosemia geraldi 118
angustella, Xanthospilopteryx perdix 19
angustifascia, Massaga 101
angustipennis, Aethodes 57
annulata, Heraclia 35
–, Xanthospilopteryx 35
ansorgei, Aegocera 69
–, Brephos 93
–, Mitrophrys 69
–, Paida 93
–, Pais 93
anthina, Aegocera 55
–, Polacanthopoda 55
Antigodasa 125, 148
Apaegocera 1, 3, 6

Arctiopais 125, 145
argentifascia, Paratuerta 108
argyrochlora, Pseudotuerta 113
–, Tuerta 113
argyrogramma, Apaegocera 7
arrosa, Rothia 130
Arrothia 125, 146
astrapeus, Phasis 74
athene suffusa, Weymeria 9
–, Veymeria 9
–, Weymeria 9
–, Xanthospilopteryx 9
atribasalis, Heraclia 29
–, Xanthospilopteryx 29
atrifusa, Heraclia 22
–, Xanthospilopteryx 22
atriventralis, Heraclia 15
–, Xanthospilopteryx 15
angustella, Xanthospilopteryx perdix 19
aurantiaca, Heraclia 157
aurantipennis, Apaegocera 7
aurea, Heraclia 35
–, Xanthospilopteryx 35

barnsi, Heraclia 34
–, Mitrophrys 69
–, Xanthospilopteryx buchholzi 34
basalifasciata, Heraclia superba 18
– separata, Xanthospilopteryx superba 18
–, Xanthospilopteryx 18
–, Xanthospilopteryx superba 18
batesi, Heraclia 34
–, Xanthospilopteryx 34
–, Xanthospilopteryx grandis 34
bicolor, Arrhotia 148
–, Arrothia bicolor 148
– bicolor, Arrothia 148
– melanobasis, Arrothia 148
bodaensis, Heraclia longipennis 31
–, Xanthospilopteryx longipennis 31
Brephos 1, 3, 90
brevivitta, Aegocera 62
–, Aegoceropsis 62
–, Aegoceropsis brevivitta 63
– brevivitta, Aegoceropsis 63
– rectilineoides, Aegocera 63
– rectilineoides, Aegoceropsis 63
– triangularis, Aegocera 62
buchholzi barnsi, Xanthospilopteryx 34
–, Eusemia 34
–, Heraclia 34
butleri adulatrix, Heraclia 16
– adulatrix, Rothia 16
– completa, Xanthospilopteryx 16
–, Eusemia 16
–, Heraclia 16
–, Rothia 16
–, Xanthospilopteryx 16

caecata, Rothia micropales 137
camilla, Paraothia 139
–, Rothia 139
campala, Heraclia aemulatrix 32
– perpallida, Xanthospilopteryx aemulatrix 32
–, Xanthospilopteryx aemulatrix 32
–, Xanthospilopteryx kirbyi 32
capensis capensis, Ovios 120
–, Eudryas 119
–, Ovios 119
–, Ovios capensis 120
– septentrionalis, Ovios 120
carminata, Rothia eriopsis 141
castaneimargo, Aegocera tigrina 54
catarhodia, Xanthospilopteryx 24
catori, Heraclia 36
–, Xanthospilopteryx 36
caudaequina, Andrhippuris 5
caudata, Agarista 46
– caudata, Hespagarista 46
– eburnea, Hespagarista 46
–, Hespagarista 46
–, Hespagarista caudata 46
celis, Arctiopais ambusta 146
Chaetostephana 4, 41
Charilina 94
Charitosemia 1, 2, 116
Choeropais 3, 97
chrysochlora, Tuerta 111, 157
– chrysochlora, Tuerta 157
– liturata, Tuerta 157
–, Tuerta chrysochlora 157
cirbyi, Heraclia 29
–, Xanthospilopteryx 29
comaria, Xanthospilopteryx 25
comorana, Aegocera 144
–, Ancarista laminifera 152
–, Nesaegocera 144
completa, Xanthospilopteryx butleri 16
confluens, Aegocera 48
–, Paraegocera 48
–, Schausia 49
congoana, Schausia leona 83
contigua, Eusemia 23
– fabricata, Rothia 54
–, Heraclia 23
–, Rothia 23
–, Xanthospilopteryx 23
coryndoni, Aegocera 84
–, Schausia 84
cosmetica, Misa 81
–, Schausia 81
–, Tuerta 81
costistrigata, Misa 81
–, Schausia 81
–, Tuerta 81
Crameria 3, 94

cruenta, Rothia 130
cyanopasta, Tuerta 112

dambuza, Schausia 157
daria, Agarista 74
–, Omphaloceps 74
–, Schausia 74
– triplagiata, Schausia 74
darocana, Aegocera 48
dayremi, Anaphela 129
–, Rothia 129
dealbata, Aegocera rectilinea 62
decora, Brephos 91
–, Noctua 91
–, Pais 91
deficiens, Eusemia 22
–, Heraclia 22
– minerva, Xanthospilopteryx 22
–, Xanthospilopteryx 22
delicia, Massaga 103
demena, Massaga 103
diffusa, Xanthospilopteryx hornimani 26
discosticta, Heraclia 37
–, Xanthospilopteryx 37
disjuncta, Metagarista maenas 52
dispar, Aegocera 76
–, Aegocera triplagiata 74
–, Epischausia 76
–, Schausia 76
–, Tuerta 76
distigma, Agarista 134
–, Rothia 134
divisa, Rothia zea 134
doenitzi, Xanthospilopteryx 31
duplicata, Xanthospilopteryx poggei 35
durbanica, Heraclia 25
–, Xanthospilopteryx 25

eburnea, Hespagarista caudata 46
echione, Agarista 45
–, Amazela 45
– funebris, Hespagarista 45
–, Hespagarista 45
– spilota, Hespagarista 45
egregia, Xanthospilopteryx 29
elegantula, Aegocera 72
elongata, Heraclia 27
–, Rothia 27
–, Xanthospilopteryx 27
eoa, Eusemia 19
–, Xanthospilopteryx 19
epiera, Rothia lasti 137
epipales, Agarista 127
–, Anaphela 128
Epischausia 4, 75
Eriopana 140
eriopis, Agarista 141
–, Anaphela 141
– carminata, Rothia 141
–, Eriopana 141
–, Ophthalmis 141
–, Pristoceraea 141
–, Rothia 141
Eudryas 2
Euphaedra 1
euphemia, Eusemia 14, 19
–, Noctua 14
–, Xanthospilopteryx 14
fabricata, Aegocera tigrina 54
–, Mitrophrys 54
–, Polyacanthopoda tigrina 54
–, Rothia contigua 54
falkensteini, Eusemia 35
fatima, Xanthospilopteryx 19
featheri, Paratuerta 106
ferrugo, Aegocera 63
–, Aegoceropsis 63
fervida, Aegocera 59
–, Aegocera fervida 60
–, Aegoceropsis 59
–, Aegoceropsis fervida 60
– affinia, Aegocera 59
– fervida, Aegocera 60
– fervida, Aegoceropsis 60
– fuscicosta, Aegocera 60
– fuscicosta, Aegoceropsis 60
– media, Aegocera 60
– media, Aegoceropsis 60
festiva, Brephos 92
–, Brephos festiva 92
– festiva, Brephos 92
– festiva, Pais 92
– nigripectus, Brephos 93
– nigripectus, Pais 93
–, Pais 92
–, Pais festiva 92
fianarantosa, Anaphela 133
–, Rothia 133
fimbria, Aegocera 60
flava, Xanthospilopteryx africana 19
–, Xanthospilopteryx geryon 14
flavifascia, Aegocera tigrina 54
flavifrons, Schausia 76
flavipennis, Heraclia 30
–, Xanthospilopteryx 30
flavisignata, Heraclia 20
–, Xanthospilopteryx 20
flaviventris, Xanthospilopteryx zenkeri 27
formosa, Noctua 65
funebris, Hespagarista echione 45
fuscicosta, Aegocera fervida 60
–, Aegoceropsis fervida 60

geometrica, Aegocera 62
–, Aegoceropsis 62
geraldi angustata, Charitosemia 118
–, Charitosemia 117, 118
–, Charitosemia geraldi 118
– geraldi, Charitosemia 118
–, Mimeusemia 118

–, Protoceraea 117
geryon flava, Xanthospilopteryx 14
–, Heraclia 14
–, Noctua 14
–, Xanthospilopteryx 14
gladiatoria, Aegocera 84
–, Schausia 84
Godasa 149, 150
gordoni, Pais 87
gracilis, Pararothia 139
–, Rothia 139
grandis, Anaphela 34
– batesi, Xanthospilopteryx 34
–, Emesia 34
–, Eusemia 34
– schoutedeni, Xanthospilopteryx 34
– subfasciata, Xanthospilopteryx 34
–, Xanthospilopteryx 34
greenawayi, Schausia 85
gruenbergi gruenbergi, Heraclia 30
–, Heraclia 29
–, Heraclia gruenbergi 30
– medjensis, Heraclia 30
– medjensis, Xanthospilopteryx 30
–, Xanthospilopteryx 29
gueneeiana, Arrothia 148
gynandra, Aegocera 68
–, Mitrophrys 68

haemaplaga, Hoplarista 86
–, Paida 86
halans, Aegocera trimeni 72
– agoma, Aegocera trimeni 72
–, Agoma trimeni 72
–, Mitrophrys 72
– transita, Aegocera trimeni 72
hampsoni, Rothia 131
haplocraspis, Pais 93
Hemituerta 1, 2, 3, 106
Heraclia 1, 2, 3, 12, 125
Hespagarista 1, 4, 44
hesparia, Massaga 101
–, Massagidia 103
hesperia, Geometra 103
–, Massaga 103
–, Massagidia 103
hollandi, Xanthospilopteryx 29
holli, Anaphela 129
–, Rothia 129
Hoplarista 3, 85
hornimani diffusa, Xanthospilopteryx 26
–, Eusemia 25
–, Heraclia 25
–, Heraclia hornimani 25
– hornimani, Heraclia 25
– hornimani, Xanthospilopteryx 25
– melanosoma, Heraclia 26
– meridionalis, Heraclia 26
– minchini, Xanthospilopteryx 26
– melanosoma, Xanthospilopteryx 26
–, Xanthospilopteryx 25
–, Xanthospilopteryx hornimani 25
hornimanni, Xanthospilopteryx 25
humfreyi, Polacanthopoda 54
humphreyi, Aegocera 54
–, Polacanthopoda 54
hypercompoides, Heraclia 26
–, Xanthospilopteryx 26
hypopyrrha, Episteme 133
–, Eusemia 133
–, Rothia 133
Hypotuerta 1, 3, 109

imperialis, Aletopus 11
inclusa, Chaetostephana 43
–, Hespagarista 43
–, Pristoceraea 43
incongruens, Eusemia 18
–, Heraclia superba 18
–, Xanthospilopteryx 18
–, Xanthospilopteryx superba 18
indecisa, Eusemia 19
–, Xanthospilopteryx 19
intercis, Charilina 95
interjecta, Hespagarista 45
interlecta, Hespagarista 45
interniplaga, Eusemia 17
–, Xanthospilopteryx 17
irangiana, Aegocera 62
isogenis, Aegocera obliqua 61
–, Aegoceropsis obliqua 61

jacksoni, Heraclia 157
jordani, Aegocera 69
joycei, Apaegocera 7
jucunda, Choeropais 98
–, Pseudospiris 98
jugans, Heraclia 16
–, Xanthospilopteryx 16
julia, Brephos 91
–, Noctua 91

karschi, Heraclia 28
– karschi, Xanthospilopteryx 29
– tristriata, Xanthospilopteryx 29
–, Xanthospilopteryx 28
–, Xanthospilopteryx karschi 29
–, Xanthospilopteryx nigridorsa 29
kenyamagaribae, Mitrophrys magna 67
kibwezensis, Aegocera 67
– aequipartita, Aegocera 67
kirbyi alice, Xanthospilopteryx 31
– campala, Xanthospilopteryx 32
–, Xanthospilopteryx 15, 28
kivuensis, Heraclia 33

ladina, Acantuerta 124
laeta, Musurgina 153

laminifera, Ancarista 151
–, Ancarista laminifera 152
– comorana, Ancarista 152
–, Euschirropterus 151
– laminifera, Ancarista 152
–, Ovios 151
–, Paratuerta 151
langazana, Schausia 157
Laritosemia 116
lasti epiera, Rothia 137
– lasti, Rothia 137
–, Rothia 137
–, Rothia lasti 137
latifascia, Leucovis 122
latreillii, Aegocera 68
–, Mitrophrys 68
leighti, Aegocera 47
leona, Aegocera 82
– congoana, Schausia 83
– leona, Schausia 83
–, Schausia 82
–, Schausia leona 83
lepta, Hylemera 122
–, Leucovis 122
Leucovis 1, 2, 121
limbomaculata, Heraclia 36
–, Xanthospilopteryx 36
littera, Heraclia 17
liturata, Tuerta chrysochlora 157
lomata, Xanthospilopteryx 28
longipennis bodaensis, Heraclia 31
– bodaensis, Xanthospilopteryx 31
–, Eusemia 31
–, Heraclia 31
–, Heraclia longipennis 31
– longipennis, Heraclia 31
–, Xanthospilopteryx 31
Lophonotidia 4, 77
lutescens, Rothia watersi 133

maculipes, Xanthospilopteryx 18
maenas, Aegocera 51
– disjuncta, Metagarista 52
– maenas, Metagarista 51
–, Metagarista 51
–, Metagarista maenas 51
magna abyssinica, Mitrophrys 67
–, Aegocera 65, 66
– kenyamagaribae, Mitrophyrs 67
– magna, Mitrophrys 67
– mesia, Mitrophrys 67
–, Mitrophrys 66
–, Mitrophrys magna 67
– obliquisigna, Mitrophrys 67
mahdi, Aegocera 108
–, Hemituerta 108
–, Tuerta 108
mantatisi, Schausia 157
mariae, Rothia 74
maritona, Massaga 101
marshalli, Paratuerta 106
martha, Rothia 137
Massaga 1, 3, 99
Massagidia 1, 4, 102
medeba, Anaphela 23
–, Eusemia 23
–, Heraclia 23
–, Xanthospilopteryx 23
media, Aegoceropsis fervida 60
–, Aegocera fervida 60
medjensis, Heraclia gruenbergi 30
–, Xanthospilopteryx 30
–, Xanthospilopteryx gruenbergi 30
melanobasis, Arrothia bicolor 148
melanochiton, Eusemia 34
melanoleuca, Lophonotidia 78
melanosoma, Heraclia hornimani 26
–, Xanthospilopteryx 26
–, Xanthospilopteryx hornimani 26
melanura, Arctiopais ambusta 146
memnonia, Misa 80
menete, Aegocera 65
– menete, Mitrophrys 66
–, Mitrophrys 65
–, Mitrophrys menete 66
–, Noctua 65
– rubida, Mitrophrys 66
meraca, Mitrophrys 74
meretrix, Eusemia 19
–, Xanthospilopteryx 19
meridionalis, Heraclia hornimani 26
mesia, Mitrophrys 67
–, Mitrophrys magna 67
Metagarista 3, 49
metagrius, Eusemia 136
–, Rothia 136
metallica, Eusemia 101
–, Massaga 101
micropales caecata, Rothia 128
– micropales, Rothia 128
–, Rothia 127
Mimeusemia 116
minchini, Xanthospilopteryx 26
–, Xanthospilopteryx hornimani 26
minerva, Xanthospilopteryx 22
–, Xanthospilopteryx dificiens 22
Misa 1, 4, 79
Mitrophrys 1, 2, 3, 64
mkabi, Schausia 157
moldaenkei, Pais 91
monslunensis, Heraclia 33
– ruwensorica, Xanthospilopteryx 33
–, Xanthospilopteryx 33
monteirona, Massaga 101
monteironis, Massaga 101
mozambica, Eusemia 37
–, Heraclia 37
–, Xanthospilopteryx 37
Musurgina 125, 152

nana, Hemituerta 108
–, Paratuerta 108
nandi, Heraclia 157
naveli, Aegocera 56
–, Polacanthopoda 56
nealces, Ovios 120
neavi, Heraclia 33
–, Xanthospilopteryx 33
Nesaegocera 125, 143, 144
niepelti, Heraclia 28
–, Heraclia zenkeri 28
–, Xanthospilopteryx 28
–, Xanthospilopteryx zenkeri 28
nigrescens, Rothia 137
nigridorsa, Eusemia 33
–, Heraclia 33
– karschi, Xanthospilopteryx 29
–, Xanthospilopteryx 33
nigrifimbriata, Rothia simyra 132
nigrimargo, Aegocera tigrina 54
nigripectus, Brephos festiva 93
–, Pais festiva 93
nigriventris, Aegocera trimeni trimeni 72
–, Xanthospilopteryx zenkeri 28
nigrobasalis, Pais 89
–, Pseudopais 89
–, Rothia 89
niveosparsa, Eusemia 20
–, Xanthospilopteryx 20
nobela, Heraclia 157
nocturna, Lophonotidia 77
noncoba, Massaga 157
norma, Aegoceropsis 59
novemmaculata, Anaphela 45
–, Eusemia 45
nugatrix, Eusemia 14
–, Xanthospilopteryx 14
nyassana, Brephos 93
–, Pais 93
nyassica, Charilina amabilis 95

obliqua, Aegocera 61
–, Aegocera obliqua 61
–, Aegoceropsis 61
–, Aegoceropsis obliqua 61
– isogenis, Aegocera 61
– isogenis, Aegoceropsis 61
– obliqua, Aegocera 61
– obliqua, Aegoceropsis 61
– remutata, Aegocera 61
– remutata, Aegoceropsis 61
obliquisigna, Aegocera 67
–, Mitrophrys 67
–, Mitrophrys magna 67
obrysos, Aegocera 142
–, Eusemia 142
–, Schausia 142
–, Schausiana 142
ochracea, Eusemia 15
–, Xanthospilopteryx 15, 28

Omphaloceps 3, 72
Ovifera, Tuerta 72
Ovios 2, 118

Paida 3, 85, 86
paidiformis, Pseudospiris 97
Pais 90
pales, Agarista 135
–, Anaphela 135
–, Eusemia 135
–, Rothia 135
pallida, Eusemia 20, 22
–, Heraclia 20
–, Xanthospilopteryx 20
pampata, Heraclia 157
panganica, Rothia 41
Paraegocera 3, 48
Pararothia 125, 138
Paratuerta 1, 2, 104
pardalina, Eusemia 15
–, Heraclia 15
–, Xanthospilopteryx 15
partita, Xanthospilopteryx poggei 35
pedasus, Agarista 135
–, Anaphela 135
–, Eusemia 135
–, Rothia 135
Pemphigostala 153
Pemphigostola 125, 153
Pentelia 3, 38
pentelia, Eusemia 39
–, Pentelia 39
perdix angustella, Xanthospilopteryx 19
–, Eusemia 19
–, Heraclia 19
–, Xanthospilopteryx 19
Perimagia 121
perpallida, Xanthospilopteryx aemulatrix campala 32
pinheyi, Pseudagoma 157
poggei duplivata, Xanthospilopteryx 35
–, Eusemia 35
–, Heraclia 35
– partita, Xanthospilopteryx 35
–, Xanthospilopteryx 35
Polacanthopoda 4, 52
postrubra, Aegoceropsis, tricolora 157
powelli, Anaphela 128
–, Rothia 128
Pristoceraea 125, 140
Protoceraea 117
Pseudagoma 157
Pseudopais 4, 88
Pseudospiris 3, 95
Pseudotuerta 2, 3, 113
pulchra, Paida 87
–, Pais 87

rectilinea, Aegocera 62
–, Aegoceropsis 62

– dealbata, Aegocera 62
rectilineoides, Aegocera 63
–, Aegocera brevivittata 63
–, Aegoceropsis brevivittata 63
rema, Tuertella 115
remutata, Aegocera obliqua 61
–, Aegoceropsis obliqua 61
rendalli angolana, Chaetostephana 43
–, Chaetostephana 43
–, Chaetostephana rendalli 43
–, Hespagarista 43
–, Metagarista 43
– rendalli, Chaetostephana 43
rhaeo, Agarista 130
– alluaudi, Rothia 130
–, Rothia 130
rhaeoides, Rothia 41, 157
Rothia 3, 40, 125, 126
rubida, Aegocera 66
–, Mitrophrys 66
–, Mitrophrys menete 66
rufodiscalis, Antigodasa 149
–, Godasa 149
–, Tuerta 149
rufoides, Xanthospilopteryx superba 18
ruspina, Aletopus 11
–, Schausia 11
–, Tuerta 11
–, Veymeria 11
ruwensorica, Xanthospilopteryx monslunensis 33

Saigonita 125
Schausia 3, 82
Schausiana 1, 125, 141
schenci, Aegocera 47
schencki, Syfanoidea 47
schoutedeni, Xanthospilopteryx grandis 34
schultzei, Misa 81
–, Tuerta 81
semialba, Aegocera trimeni trimeni 72
separata, Aegoceropsis 84
–, Xanthospilopteryx superba basalifasciata 18
septentrionalis, Ovios 120
–, Ovios capensis 120
signata, Xanthospilopteryx 33
simplex, Rothia 135
simyra, Anaphela 132
– nigrifimbriata, Rothia 132
–, Rothia 132
sinefascia, Rothia tranquilla XXX
smithii, Xanthospilopteryx 17
spilota, Hespagarista echione 45
subfasciata, Xanthospilopteryx grandis 34
sublaeta, Brephos 157
subterminalis, Rothia 89
suda, Heraclia aemulatrix 32
–, Xanthospilopteryx aemulatrix 32
suffusa, Weymeria athene 9
superba basalifasciata, Heraclia 18
– basalifasciata separata, Xanthospilopteryx 18
– basalifasciata, Xanthospilopteryx 18
–, Eusemia 17
–, Heraclia 17
–, Heraclia superba 18
– incongruens, Heraclia 18
– incongruens, Xanthospilopeteryx 18
– rufoides, Xanthospilopteryx 18
– superba, Heraclia 18
– superba, Xanthospilopteryx 18
– ugandana, Heraclia 18
– ugandana, Xanthospilopteryx 18
–, Xanthospilopteryx 17
–, Xanthospilopteryx superba 18
Syfanoidea 1, 3, 46
synemonistis, Pemphigostola 155

tanganyikana, Xanthospilopteryx 18
tenuifascia, Massagidia 104
tenuis, Rothia tranquilla 128
terminatis, Anaphela 21
–, Eusemia 21
–, Heraclia 21
–, Xanthospilopteryx 21
tessmani, Xanthospilopteryx 29
thomensis, Acantuerta 124
–, Aegocera 124
–, Tuerta 124
thruppi, Eusemia 24
–, Heraclia 24
–, Rothia 24
–, Xanthospilopteryx 24
tigrina, Aegocera 54
–, Aegocera tigrina 54
– albifascia, Aegocera 54
– castaneimargo, Aegocera 54
– fabricata, Aegocera 54
– fabrioata, Polacanthopoda 54
– flavifascia, Aegocera 54
–, Hespagarista 53
–, Metagarista 53
– nigrimargo, Aegocera 54
–, Polacanthopoda 53
–, Polacanthopoda tigrina 54
– tigrina, Aegocera 54
– tigrina, Polacanthopoda 54
–, Xanthospilopteryx 53
tranquilla, Anaphela 128
–, Eusemia 127
–, Rothia 128
– sinefascia, Rothia 136
– tenuis, Rothia 128
transiens, Hypotuerta 110
–, Tuerta 110
–, Xanthospilopteryx aisha 24
transita, Aegocera trimeni halans 72
triangularis, Aegocera brevivittata 62
–, Eusemia 74

zoologique en question sans pour cela déborder le cadré de l'ensemble de l'ouvrage. Sont invités à collaborer, tous les auteurs qui désireraient publier des révisions systématiques de groupes zoologiques récents. Conformément aux voeux des auteurs, leurs traités seront imprimés en anglais, en français et en allemand. Les honoraires des auteurs sont fixés à DM 55,— par placard de l'imprimerie (= 16 pages) et 5 exemplaires gratuits après parution de leurs traités. Pour plus amples détails, se reporter à les livraisons No. 72 et suivantes qui vient de paraître depuis 1958. Les éditeurs fournissent tous renseignements sur demande ou des indications précises à l'intention des auteurs. Dans la série ,,DAS TIERREICH", on publiera aussi des révisions sommaires de groupes zoologiques comme ,,Checklists".

Das ,,TIERREICH" wurde von der Deutschen Zoologischen Gesellschaft in der Absicht gegründet, eine vollständige systematische Darstellung aller rezenten Gruppen des Tierreichs zu schaffen. Nach dem zweiten Weltkrieg wurde das Unternehmen abgebrochen, da das Ziel der Vollständigkeit unerreichbar schien. Damit entfielen jedoch Anreiz und Möglichkeit, umfassende Revisionen größerer Tiergruppen in fortlaufenden Lieferungen zu veröffentlichen. Aus diesem Grunde hat sich der Verlag Walter de Gruyter & Co. (Berlin) entschlossen, das ,,TIERREICH" zwar grundsätzlich unter den ursprünglichen Gesichtspunkten, darüber hinaus aber lediglich als eine zwanglose Reihenfolge systematischer Monographien wieder aufleben zu lassen, ohne zu beabsichtigen, alle zoologischen Gruppen nach einem feststehenden Plan abzuhandeln. Diese Monographien sollen sich nicht ausschließlich auf die Belange der Taxionomie beschränken, sondern auch anatomische, biologische, physiologische und andere Tatsachen berücksichtigen, sofern sie geeignet sind, die betreffende Tiergruppe näher zu charakterisieren, ohne den Rahmen des Gesamtwerkes zu sprengen. Zur Mitarbeit sind alle Autoren eingeladen, die den Wunsch haben, systematische Revisionen rezenter Tiergruppen zu veröffentlichen. Entsprechend den Wünschen der Verfasser werden ihre Abhandlungen in Englisch, Französisch und Deutsch gedruckt. Die Autoren erhalten ein Honorar in Höhe von DM 55,— für jeden Druckbogen (= 16 Seiten) und 5 Sonderdrucke nach Erscheinen ihrer Beiträge. Weitere Einzelheiten mögen den seit 1958 erschienenen Lieferungen Nr. 72 und folgenden entnommen werden. Auf Wunsch erteilen die Herausgeber jegliche Auskünfte oder geben genaue Anweisungen an die Autoren ab. In der Reihe ,,DAS TIERREICH" erscheinen auch summarische Bearbeitungen von größeren Tiergruppen in Form sogenannter ,,Checklists".

Die Herausgeber

†Prof. Dr. Robert Mertens (Wirbeltiere)
Natur-Museum und Forschungs-Institut Senckenberg
Senckenberg-Anlage 25, 6000 Frankfurt am Main 1

Prof. Dr. Willi Hennig (Wirbellose)
Staatliches Museum für Naturkunde
Arsenalplatz 3, 7140 Ludwigsburg

Dr. Heinz Wermuth (Schriftleiter)
Staatliches Museum für Naturkunde
Arsenalplatz 3, 7140 Ludwigsburg

www.ingramcontent.com/pod-product-compliance
Lightning Source LLC
LaVergne TN
LVHW042243300926
849161LV00003B/448
* 9 7 8 3 1 1 2 4 2 0 9 5 9 *